Scala
开发快速入门

周志湖 牛亚真 编著

清华大学出版社
北京

内 容 简 介

本书以 Scala 语言的核心语法为主线,通过大量实例代码循序渐进地介绍了 Scala 语言的语法特性。第 1 章介绍 Scala 语言及开发环境的安装、Intellij IDEA 的使用、Scala 语言的交互式命令行。第 2~3 章重点介绍 Scala 语言的基础语法结构。第 4 章,重点介绍 Scala 语言重要的数据结构 collection(集合)。第 5 章,重点介绍 Scala 语言中面向函数编程的语法特性。第 6~8 章,重点介绍 Scala 语言面向对象编程的语法特性。第 9 章,介绍 Scala 语言中的模式匹配及原理。第 10 章,详细介绍 Scala 语言的类型系统。第 11 章,对 Scala 语言中的隐式转换及原理进行详细介绍。第 12 章,重点介绍 Scala 中的并发编程。第 13 章,介绍 Java 语言与 Scala 语言间的互操作。

本书还提供了所有实例的源代码与素材文件,供读者上机练习使用,读者可从网上下载本书资源文件。

本书适用于 Scala 语言初学者、爱好者,大数据开发人员,软件培训机构以及计算机专业的学生使用。

本书封面贴有清华大学出版社防伪标签,无标签者不得销售。
版权所有,侵权必究。侵权举报电话: 010-62782989 13701121933

图书在版编目(CIP)数据

Scala 开发快速入门 / 周志湖,牛亚真编著.-北京: 清华大学出版社,2016
ISBN 978-7-302-44413-8

I. ①S… II. ①周… ②牛… III. ①JAVA 语言—程序设计 IV. ①TP312

中国版本图书馆 CIP 数据核字(2016)第 168661 号

责任编辑:王金柱
封面设计:王 翔
责任校对:闫秀华
责任印制:王静怡

出版发行:清华大学出版社
网　　址:http://www.tup.com.cn,http://www.wqbook.com
地　　址:北京清华大学学研大厦 A 座　　邮　编:100084
社 总 机:010-62770175　　邮　购:010-62786544
投稿与读者服务:010-62776969,c-service@tup.tsinghua.edu.cn
质 量 反 馈:010-62772015,zhiliang@tup.tsinghua.edu.cn

印 装 者:三河市金元印装有限公司
经　　销:全国新华书店
开　　本:190mm×260mm　　印　张:19.75　　字　数:506 千字
版　　次:2016 年 9 月第 1 版　　印　次:2016 年 9 月第 1 次印刷
印　　数:1~3000
定　　价:59.00 元

产品编号:069160-01

推 荐 序

蒲晓蓉 博士

经过近两三年的媒体炒作、技术发展及部分行业应用落地，大数据不再悬在空中，而是真正地在影响着你我的日常生活，"数据就是资源"、"数据就是石油"、"数据就是金矿"等数据价值理念已深入人心。

IDC 出版的数字世界研究报告显示，2013 年人类产生、复制和消费的数据量达到 4.4ZB，而到 2020 年，数据量将增长 10 倍，达到 44ZB，中国众多的人口数量及广泛应用的硬件基础设施将使我国 2020 年的数据量增长 24 倍，为我国大数据产业的发展提供良好的基础。党和政府为适应经济发展的新常态，稳步推进经济结构调整，助力经济转型升级，2015 年 10 月，党的十八届五中全会公报提出要实施"国家大数据战略"，大数据第一次写入党的全会决议，从此开启了我国大数据建设的新篇章。

回归现实，大数据行业面临着诸如人才稀缺、技术和创业门槛高、行业可借鉴经验少等众多问题，其中人才稀缺是当前最迫切需要解决的问题。虽然基于开源 Hadoop 生态体系的大数据解决方案一定程度上降低了技术门槛和创业门槛，但面临具体应用场景时，开源体系中可选择的工具众多，如 SQL-On-Hadoop 工具就有 Hive、Impala、Presto、Stinger 等，这对技术人员的要求非常高，因为不同的工具开发语言、兼容性等可能都不同。Apache Spark 的出现极大地降低了大数据从业人员的门槛，它"一个软件栈内完成所有大数据分析任务（one stack to rule them all）"的思想及基于内存的计算模式一定程度上代表着未来大数据技术的发展方向。

Scala 语言具有简洁、优雅、与 Java 语言的良好互操作性及融合函数式与面向对象编程思想等诸多优点，值得每个优秀的程序员花时间去学习。在大数据领域，除 Apache Spark 底层采用 Scala 语言实现外，还有 Apache Kafka 及 Apache Gearpump 等优秀的大数据处理框架；在其他应用开发领域，还有 Akka 并发编程框架、Lagom 微服务框架、Play Web 应用开发框架等优秀的产品，这已足以证明该语言的强大。

本书作者具备深厚的计算机专业背景、科研能力和写作水平，对新事物充满好奇心，善于学习新知识。通过自学 Scala 语言，积累了丰富的第一手资料和经验，并将个人学习经验通过本书分享出来，有助于 Scala 语言初学者的学习，一定程度上推动国内 Scala 语言的普及。

蒲晓蓉
2016 年 6 月

蒲晓蓉，博士，电子科技大学计算机科学与工程学院教授，英才实验学院副院长。全国计算机学会高级会员；第二届机械工业教育协会会员；电子工业出版社特聘专家；教育部高校教师培训中心特聘专家。出版国家级"十五"、"十一五"规划教材等8部，《计算机操作系统》国家级资源共享课负责人，获四川省教学成果二等奖、四川省科技进步二等奖。主要研究方向为机器学习、计算机视觉和操作系统。

前　言

为什么要写这本书

　　Apache Spark、Apache Kafka 等基于 Scala 语言实现的框架在大数据生态圈内占有举足轻重的地位，它们一定程度上引领着大数据最前沿的技术方向，另外 Akka、Apache Gearpump 等基于 Scala 语言实现的框架在大数据生态圈内也备受关注，工业实践已经证明了 Scala 语言的强大，这使 Scala 语言近一两年名声大噪，进而引起众多编程语言爱好者及各大公司的亲睐，国内外也因此掀起了一场学习 Scala 语言的热潮。

　　Scala 语言基于 JVM 平台，能够与 Java 语言进行良好地互操作，它最大的特点是可以将面向对象编程语言与函数式编程语言的特点结合起来。Scala 语言的简洁性和表达能力，使得 Scala 语言与 Java 语言相比，同样的功能用 Scala 实现代码量要少很多。Scala 语言的众多优点，让许多业内名家对其推崇备至，这其中便有 Java 之父 James Gosling 和 Groovy 语言创始人 James Strachan。

　　"If I were to pick a language to use today other than Java, it would be Scala."[1]（如果现在让我选择使用除 Java 之外的一门语言的话，那一定是 Scala）——Java 之父 James Gosling

　　"Though my tip though for the long term replacement of javac is Scala. I'm very impressed with it! I can honestly say if someone had shown me the Programming in Scala book by Martin Odersky, Lex Spoon & Bill Venners back in 2003 I'd probably have never created Groovy."（我认为将来可能替代 javac 的就是 Scala，它极大地震撼了我，老实说如果有人在 2003 年把 Martin Odersky、Lex Spoon 和 Bill Venners 写的那本《Programming in Scala》书拿给我看了的话，那我可能就不会再去发明 Groovy 语言了）[2]——Groovy（另一种基于 JVM 平台的语言）创始人 James Strachan。

　　目前国内 Scala 方面教材的短缺一定程度上限制了 Scala 语言在国内的普及，本书试图弥补这一空白。

本书内容

　　本书是笔者在 Scala 学习、工作实践及培训过程中的心得体会和系统总结。内容涵盖 Scala 语言基础知识，包括变量的定义、程序控制结构、Scala 集合操作；Scala 语言中级知识，包括 Scala 函数式编程、Scala 面向对象编程、Scala 模式匹配；Scala 语言高级编程知识，包括 Scala 类型参数、隐式转换、Scala 并发编程、Scala 与 Java 互操作。书中利用大量的具体示

---

[1] http://www.edureka.co/blog/why-scala-is-getting-popular/
[2] http://macstrac.blogspot.com/2009/04/scala-as-long-term-replacement-for.html

例和实际案例来说明 Scala 语言的应用，既能够掌握核心语法，又能够理解其背后的深层次原理。

读者对象

（1）Spark、Kafka 等框架二次开发人员

Spark、Kafka 等大数据处理框架目前在稳定性、扩展性方面虽然已经取得了长足的进步，但在实际使用时难免遇到问题，此时需要深入到内核源代码中分析问题，在理解其架构原理与实现细节的基础上通过修改内核源代码来解决问题，这需要开发人员有很强的 Scala 语言功底。

（2）Spark、Kafka 等框架应用开发人员

在学习 Spark、Kafka 等 Scala 语言实现的大数据技术框架时，这些框架大多都提供了 Java、Scala、Python 等上层应用 API 编程接口，但仅仅学会使用其上层 API 是不够的，因为上层 API 抽象程度较高，各框架的底层实现细节、设计原理等对开发人员来说是一个黑匣子，在遇到性能等问题时，开发人员如果对底层实现细节不熟悉的话可能很难进行程序的性能优化。

（3）Spark、Kafka 等框架运维工程师

目前 Spark、Kafka 等大数据处理框架基本上已经成为各大公司大数据解决方案的标配，但基于开源解决方案常常会面临一些技术风险，框架本身的问题及实际应用场景的不同可能会遇到很多故障，因此对于这些使用 Scala 语言实现的框架而言，运维工程师也需要知晓各框架的底层细节，这样才能够迅速定位问题并找到合适的解决办法。

（4）从事大数据技术的 Java 开发人员

Scala 语言完全兼容 Java 语言，Scala 语言中大量使用了 Java 语言现有的库，因此如果你是一个 Java 大数据开发人员，在此建议学习 Scala 语言，相信它会给你带来全新的感受，甚至有抛弃 Java 的念头。

（5）大数据技术开源爱好者

Spark、Kafka 等是大数据技术框架中的明星，Scala 语言已经通过了工业界的检验，学习 Scala 语言为学习这些框架的设计原理打下坚实的基础，为以后参与开源、学习工业界最先进的大数据技术架构的优秀思想打下了坚实的基础。

如何阅读本书

本书涵盖三大主要部分：

（1）Scala 语言基础篇，主要介绍 Scala 语言入门基础知识、变量的定义、程序控制结构、Scala 集合。

（2）Scala 语言中级篇，主要介绍 Scala 函数式编程、Scala 面向对象编程、Scala 模式匹配。

（3）Scala 语言高级篇，主要介绍 Scala 类型参数、隐式转换、Scala 并发编程、Scala

与Java互操作。

如果你是一名初学者，请按照书的顺序从第一章基础理论部分开始学习，学习时先认真看书中的代码示例和说明，然后照着代码亲自动手实践，这样可以达到事半功倍的学习效果；如果你是一名Scala资深用户，你可以自由阅读各章内容，相信书中部分内容肯定能够给你带来一些新的启发。

资源下载

本书代码可以从 http://pan.baidu.com/s/1nvyDC1r（注意数字和字母大小写）下载。如果下载有问题，请电子邮件联系 booksaga@163.com，邮件标题为"Scala开发快速入门-代码"。

勘误和支持

由于笔者水平有限，加之编写时间仓促，书中难免出现一些错误或不准确的地方，恳请读者批评指正。有任何问题，可以关注微信公众号：ScalaLearning 或者通过邮箱 403408607@qq.com 联系，编者将尽最大努力及时为读者提供相应的解答。

编　者

2016年6月于杭州

致　谢

感谢我的导师蒲晓蓉教授，在我攻读硕士研究生时，是她给予我项目、学术及为人处理事方面的诸多指导，是蒲老师带我进入机器学习、计算机视觉的世界，从而有幸能够接触到 Scala 语言。

感谢清华大学出版社的王金柱老师及他的团队，王老师在近八个月的时间内始终支持我的写作，是他的鼓励和帮助让我能够在繁忙的工作之余孜孜不辍地完成本书的编写。

感谢对本书部分章节提出建议的师兄杨智杰，感谢部门总经理陈霄一直以来对我的信任和工作上的指导，还要感谢赵永标、秦海龙在工作上对我的支持与帮助。

最后，感谢父母及亲朋好友的鼓励和支持，感谢他们在我困顿中给予无私帮助；感谢老婆对我生活的悉心照顾和宽容；感谢我可爱的女儿逸霖，是她的笑容让我忘却烦恼，坚持每天写作。

谨以此书献给我最亲爱的家人和朋友，以及所有热爱 Scala 语言及开源技术的朋友们。

周志湖
2016 年 6 月于杭州

目 录

第 1 章 Scala 入门 ... 1

1.1 Scala 简介 ... 2
1.2 Scala 开发环境搭建 ... 2
- 1.2.1 软件准备 ... 2
- 1.2.2 JDK 的安装与配置 ... 3
- 1.2.3 Scala SDK 的安装与配置 ... 5
- 1.2.4 Intellij IDEA 的安装与配置 ... 6

1.3 Scala Hello World ... 8
- 1.3.1 创建 Scala Project ... 8
- 1.3.2 配置项目代码目录结构 ... 9
- 1.3.3 创建应用程序对象 ... 11
- 1.3.4 运行代码 ... 12

1.4 Intellij IDEA 常用快捷键 ... 13
- 1.4.1 代码编辑类常用快捷键 ... 13
- 1.4.2 导航快捷键 ... 14
- 1.4.3 编译、运行及调试 ... 15
- 1.4.4 代码格式化 ... 15

1.5 交互式命令行使用 ... 15
- 1.5.1 Scala 内置交互式命令行 ... 15
- 1.5.2 Scala Console 交互式命令行 ... 17
- 1.5.3 Scala Worksheet 的使用 ... 19

小 结 ... 20

第 2 章 变量及基本数据类型 ... 21

2.1 变 量 ... 22

2.1.1　变量定义 ··· 22
　　　2.1.2　lazy 变量 ··· 23
　2.2　基本数据类型 ··· 24
　　　2.2.1　Int 类型 ··· 24
　　　2.2.2　Float 类型 ·· 25
　　　2.2.3　Double 类型 ·· 25
　　　2.2.4　Char 类型 ··· 25
　　　2.2.5　String 类型 ··· 26
　　　2.2.6　Boolean 类型 ··· 27
　2.3　基本类型操作 ··· 27
　　　2.3.1　算术运算操作 ··· 27
　　　2.3.2　关系运算操作 ··· 28
　　　2.3.3　逻辑运算操作 ··· 28
　　　2.3.4　位运算操作 ·· 29
　　　2.3.5　对象比较运算操作 ··· 30
　　　2.3.6　字符串运算操作 ·· 31
　2.4　运算符的优先级 ·· 32
　2.5　元组类型 ··· 33
　2.6　符号类型 ··· 34
　小　　结 ·· 35

第 3 章　程序控制结构 ··· 36

　3.1　if 语句 ·· 37
　　　3.1.1　if 语句 ··· 37
　　　3.1.2　if…else…语句 ··· 37
　　　3.1.3　if…else if…else…语句 ·· 37
　　　3.1.4　if 的嵌套使用 ·· 38
　　　3.1.5　if 表达式 ·· 39
　3.2　while 循环语句 ··· 39
　　　3.2.1　while 语句的使用 ··· 39
　　　3.2.2　do while 语句的使用 ··· 40
　3.3　for 循环语句 ·· 41
　　　3.3.1　基础 for 循环 ·· 41
　　　3.3.2　有过滤条件的 for 循环 ·· 44

3.3.3　多重 for 循环 ··· 44

　　　3.3.4　作为表达式的 for 循环 ··· 45

　小　结 ·· 45

第 4 章　集　合··· 46

　4.1　集合简介 ··· 47

　4.2　数　组 ·· 49

　　　4.2.1　定长数组 ·· 49

　　　4.2.2　变长数组 ArrayBuffer ··· 50

　　　4.2.3　数组的遍历 ··· 52

　　　4.2.4　遍历生成数组 ·· 53

　　　4.2.5　常用函数 ·· 53

　　　4.3.6　多维数组 ·· 54

　4.3　列　表（List）·· 55

　　　4.3.1　列表的创建 ··· 55

　　　4.3.2　List 常用函数 ·· 56

　　　4.3.3　List 伴生对象方法 ·· 58

　4.4　集（Set）··· 59

　4.5　映　射（Map）··· 61

　4.6　队　列（Queue）·· 63

　4.7　栈（Stack）··· 64

　小　结 ·· 66

第 5 章　函　数··· 67

　5.1　函　数 ·· 68

　5.2　值 函 数 ·· 69

　　　5.2.1　值函数的定义 ·· 69

　　　5.2.2　值函数的简化 ·· 71

　5.3　高阶函数 ··· 73

　　　5.3.1　高阶函数的定义 ··· 73

　　　5.3.2　高阶函数的使用 ··· 74

　5.4　闭　包 ·· 78

　5.5　函数柯里化 ·· 80

　5.6　部分应用函数 ··· 81

5.7 偏函数 83
小　结 85

第 6 章　Scala 面向对象编程（上） 86

6.1 类与对象 87
6.1.1 类的定义 87
6.1.2 创建对象 88
6.1.3 类成员的访问 88
6.1.4 单例对象 90
6.1.5 应用程序对象 91
6.1.6 伴生对象与伴生类 93

6.2 主构造函数 96
6.2.1 主构造函数的定义 96
6.2.2 默认参数的主构造函数 98
6.2.3 私有主构造函数 99

6.3 辅助构造函数 100
6.3.1 辅助构造函数的定义 100
6.3.2 辅助构造函数中的默认参数 102

6.4 继承与多态 104
6.4.1 类的继承 104
6.4.2 构造函数执行顺序 107
6.4.3 方法重写 108
6.4.4 多态 109

6.5 成员访问控制 111
6.5.1 默认访问控制 112
6.5.2 protected 访问控制 113
6.5.3 private 访问控制 114
6.5.4 private[this] 访问控制 116
6.5.5 主构造函数中的成员访问控制 121

6.6 抽象类 125
6.6.1 抽象类的定义 125
6.6.2 抽象类的使用 126

6.7 内部类与内部对象 128

6.8 匿名类 130

小 结 131

第 7 章 Scala 面向对象编程（下） 132

7.1 trait 简介 133
7.2 trait 的使用 135
    7.2.1 trait 的几种不同用法 135
    7.2.2 混入 trait 的类对象构造 138
    7.2.3 提前定义与懒加载 140
7.3 trait 与类 144
    7.3.1 trait 与类的相似点 144
    7.3.1 trait 与类的不同点 146
7.4 多重继承问题 147
7.5 自身类型 148
小 结 151

第 8 章 包（package） 152

8.1 包的定义 153
8.2 包的使用和作用域 155
    8.2.1 包的使用 155
    8.2.2 包作用域 156
8.3 包 对 象 159
8.4 import 高级特性 160
    8.4.1 隐式引入 160
    8.4.2 引入重命名 160
    8.4.3 类隐藏 161
小 结 162

第 9 章 模式匹配 163

9.1 模式匹配简介 164
9.2 模式匹配的 7 大类型 167
    9.2.1 常量模式 167
    9.2.2 变量模式 167
    9.2.3 构造函数模式 169
    9.2.4 序列模式 170

   9.2.5 元组模式 ·········· 171
   9.2.6 类型模式 ·········· 172
   9.2.7 变量绑定模式 ·········· 173
  9.3 模式匹配原理 ·········· 174
   9.3.1 构造函数模式匹配原理 ·········· 174
   9.3.2 序列模式匹配原理 ·········· 176
  9.4 正则表达式与模式匹配 ·········· 177
   9.4.1 Scala 正则表达式 ·········· 177
   9.4.2 正则表达式在模式匹配中的应用 ·········· 180
  9.5 for 循环中的模式匹配 ·········· 183
  9.6 模式匹配与样例类、样例对象 ·········· 185
   9.6.1 模式匹配与样例类 ·········· 185
   9.6.2 模式匹配与样例对象 ·········· 187
  小 结 ·········· 190

第 10 章　隐式转换 ·········· 191

  10.1 隐式转换简介 ·········· 192
  10.2 隐式转换函数 ·········· 193
   10.2.1 隐式转换函数的定义 ·········· 193
   10.2.2 隐式转换函数名称 ·········· 194
  10.3 隐式类与隐式对象 ·········· 195
   10.3.1 隐式类 ·········· 195
   10.3.2 隐式对象 ·········· 196
  10.4 隐式参数与隐式值 ·········· 197
   10.4.1 隐式参数 ·········· 197
   10.4.2 隐式值 ·········· 199
   10.4.3 隐式参数使用常见问题 ·········· 201
  10.5 隐式转换规则与问题 ·········· 204
   10.5.1 隐式转换的若干规则 ·········· 204
   10.5.2 隐式转换需注意的问题 ·········· 208
  小 结 ·········· 210

第 11 章　类型参数 ·········· 211

  11.1 类与类型 ·········· 212

11.2 泛型 ······ 214
11.2.1 泛型类 ······ 215
11.2.2 泛型接口与泛型方法 ······ 216
11.2.3 类型通配符 ······ 217
11.3 类型变量界定 ······ 220
11.4 视图界定 ······ 222
11.5 上下文界定 ······ 223
11.6 多重界定 ······ 226
11.7 协变与逆变 ······ 227
11.8 高级类型 ······ 230
11.8.1 单例类型 ······ 231
11.8.2 类型投影 ······ 235
11.8.3 类型别名 ······ 237
11.8.4 抽象类型 ······ 238
11.8.5 复合类型 ······ 240
11.8.6 函数类型 ······ 241
小 结 ······ 243

第 12 章 Scala 并发编程基础 ······ 244
12.1 Scala 并发编程简介 ······ 245
12.1.1 重要概念 ······ 245
12.1.2 Actor 模型 ······ 247
12.1.3 Akka 并发编程框架 ······ 247
12.2 Actor ······ 249
12.2.1 定义 Actor ······ 249
12.2.2 创建 Actor ······ 250
12.2.3 消息处理 ······ 255
12.2.4 Actor 的其他常用方法 ······ 260
12.2.5 停止 Actor ······ 261
12.3 Typed Actor ······ 265
12.3.1 Typed Actor 定义 ······ 265
12.3.2 创建 Typed Actor ······ 266
12.3.3 消息发送 ······ 267
12.3.4 停止运行 Typed Actor ······ 269

12.4 Dispatcher ··· 271
　　12.4.1　常用 Dispatcher ··· 271
　　12.4.2　ExecutionService ·· 274
12.5 Router ·· 277
12.6 容　错 ·· 279
　　12.6.1　Actor 的 4 种容错机制 ·· 279
　　12.6.2　Supervison ··· 279
小　结 ·· 285

第 13 章　Scala 与 Java 的互操作 ·· 286

13.1 Java 与 Scala 集合互操作 ·· 287
　　13.1.1　Java 调用 Scala 集合 ··· 287
　　13.1.2　Scala 调用 Java 集合 ··· 288
　　13.1.3　Scala 与 Java 集合间相互转换分析 ································ 289
13.2 Scala 与 Java 泛型互操作 ·· 290
　　13.2.1　Scala 中使用 Java 泛型 ·· 291
　　13.2.2　Java 中使用 Scala 泛型 ·· 291
13.3 Scala trait 在 Java 中的使用 ·· 293
13.4 Scala 与 Java 异常处理互操作 ·· 298
小　结 ·· 299

参考文献 ··· 300

Scala 入门

本章主要介绍 Scala 语言的特点、Scala 开发环境搭建，如何进行 Hello World 应用程序的编写、Intelli IDEA 常用快捷键的使用及 Scala 交互式命令行的使用。

1.1 Scala 简介

　　Scala 语言（Scalable Lanaguage）是集面向对象编程思想与函数式编程思想于一身的通用编程语言，由 Martin Odersky 教授及其领导的瑞士洛桑联邦高等理工学院程序方法实验室团队于 2001 年发明，然后于 2004 年对外发布，起初只能在 Java 虚拟机（JVM）上运行，随后团队又将 Scala 语言扩展到.Net 平台上。目前 Scala 语言在.Net 平台上运行仍然不够稳定，主要原因是 Scala 大量重用了 Java 语言提供的库，在.Net 平台上使用时会存在不少问题，本书后续章节如果没有特别指出，描述的内容全部都是基于 JVM 平台的。

　　Scala 语言在创立之初，并没有引起太多的关注，随着 Apache Spark、Apache Kafka 及 Apache Gearpump 等大数据处理框架的出现及爆发式增长，Scala 语言在大数据处理、分布式计算等领域受到广泛关注。虽然这些大数据框架提供了多语言编程接口，但毕竟 Scala 语言是这些框架的原生开发语言，特别是对于那些想学习这些框架内核原理的工程师们来说，Scala 语言已经成为大数据工程师们必备的编程语言之一。

　　为什么 Apache Spark、Apache Kafka 及 Gearpump 等大数据处理框架要使用 Scala 语言作为其原生开发语言呢？这是因为 Scala 语言具有以下特点。

　　（1）能够无缝集成 Java 语言：Scala 语言能够与 Java 良好地进行互操作，事实上 Scala 语言可以看作是 Java++，它完全兼容 Java 语言同时又提供了很多新特性。

　　（2）纯面向对象编程语言：Scala 语言具有统一的对象模型，所有的值都是对象，而所有的操作都由消息发送完成。

　　（3）函数式编程语言：Scala 语言也是函数式编程语言，提供了高阶函数（Higher-order functions）、闭包（Closures）、模式匹配（Pattern matching）、单一赋值（Single assignment）、延迟计算（Lazy evaluation）、类型推导（Type inference）及尾部调用优化（Tail call optimization）等多种语法功能。

　　由于 Scala 语言的这些特点，与 Java 语言编写的代码相比，Scala 的代码量要少很多。

1.2 Scala 开发环境搭建

1.2.1 软件准备

　　Scala 开发环境所需的软件如表 1-1 所示。

表 1-1 Scala 开发软件需求

软件名称	版本	用途与下载地址
JDK	1.7.0_79	Java 语言的开发工具包，下载地址： http://www.oracle.com/technetwork/java/javase/downloads/jdk7-downloads-1880260.html

（续表）

软件名称	版本	用途与下载地址
Scala SDK	2.10.4	Scala 语言的开发工具包，下载地址：http://www.scala-lang.org/download/2.10.4.html
IntelliJ IDEA	15.0.4	集成开发环境，用于进行 Scala 大型项目开发，下载地址：http://www.jetbrains.com/idea/download

1.2.2 JDK 的安装与配置

运行"jdk1.7.0_79.msi"后，可以看到图 1-1 所示的界面。

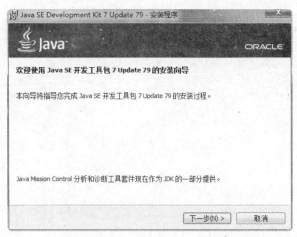

图 1-1　JDK 安装启动界面

单击"下一步"按钮，可以看到图 1-2 所示的界面，"开发工具"和"源代码"都选择安装，然后将安装目录设置为"C:\Java"。

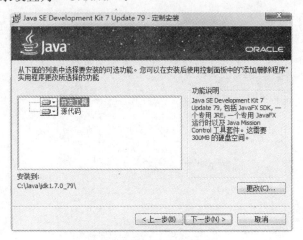

图 1-2　定制安装界面

完成后单击"下一步"按钮，JDK 会自动完成安装，完成后会弹出安装 JRE 提示，同样将其安装在"C:\Java"目录下即可。

配置环境变量，右键单击"计算机"，选择"属性"，如图 1-3 所示。

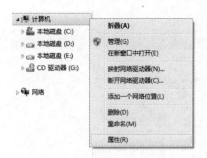

图 1-3 系统配置页面

在"控制面板主页"单击"高级系统设置",进入"系统属性"对话框,如图 1-4 所示。

图 1-4 "系统属性"对话框

在"高级"选项卡中单击"环境变量",进入"环境变量"配置对话框,在系统变量中单击"新建",设置变量名为 JAVA_HOME,设置变量值为 C:\Java\jdk1.7.0_79,然后单击"确定"按钮完成设置。

图 1-5 设置 JAVA_HOME 环境变量

在"系统变量"中查找名称为 Path 的环境变量,单击"编辑",在变量中添加

";%JAVA_HOME%\bin",然后单击"确定"按钮,如图1-6所示。

图1-6 添加bin目录到环境变量Path中

最后在"系统变量"中,单击"新建"按钮,设置变量名为"CLASS_PATH"、变量值为".;%JAVA_HOME%\lib",再单击"确定"按钮,至此完成了整个JDK的安装与配置。打开DOS,输入java –version命令,可以看到图1-7所示的输出结果,则说明JDK安装成功。

图1-7 java –version命令输出结果

1.2.3 Scala SDK的安装与配置

单击运行"scala-2.10.4.msi",出现如图1-8所示的安装启动界面。

图1-8 Scala安装启动界面

单击"next"按钮,选择同意接受其"License Agreement"条款并单击"next"按钮,进入图 1-9 所示的 Scala 安装定制界面,选择安装所有的组件并设置安装目录为"C:\scala"。

图 1-9　Scala 安装定制界面

单击"next"按钮,再单击"install"按钮,程序自动完成安装并将 bin 目录添加到系统环境变量中。打开 DOS,输入 scala –version 命令,如果有图 1-10 所示的输出结果,则表示 Scala 安装成功。

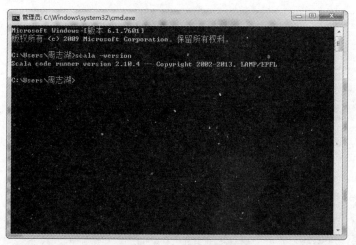

图 1-10　scala –version 命令输出结果

1.2.4　Intellij IDEA 的安装与配置

Intellij IDEA 15.0.4 社区版直接按默认条件安装即可,该版本的 Intellij IDEA 已经内置了 Scala 插件无需再手动安装。为方便后期讲解,这里先对 Intellij IDEA 整体界面[1]进行完整的介绍,如图 1-11 所示。

1. 菜单和工具栏(Menus and toolbars):包含项目需要使用的大多数命令。
2. 导航栏(Navigation bar):显示当前项目及正在编辑的文件。

---

[1] https://www.jetbrains.com/idea/help/quick-start-guide.html

3. 状态栏（The status bar）：显示项目、整个开发环境的状态，输出关于项目的警告及其他消息。

4. 编辑器（Editor）：用于创建和修改代码。

5. 工具窗口（Tool windows）：具备多种功能，包括查看项目的文件结构、编译执行程序等。

6. 左边界（Left gutter）：用于设置代码断点、显示代码行号等。

7. 右边界（Right gutter）：用于监测代码质量，包括显示代码出错、警告等信息，右上角会显示整个文件代码的质量情况。

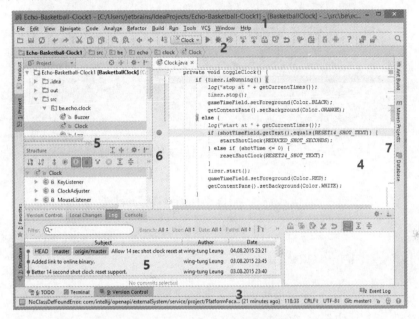

图 1-11　Intellij IDEA 整体布局

对于开发人员来说，最喜爱的使用风格是 Darcula，在 File | Settings | Appearance & Behaviour | Appearance 中进行设置，如图 1-12 所示，将 Theme 设置为"Darcula"即可。

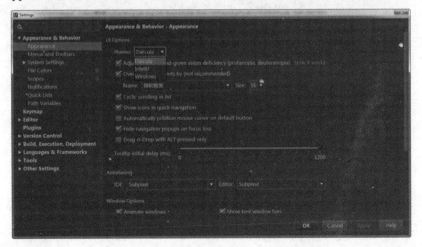

图 1-12　设置 Intellij IDEA 主题风格

不过，在正式使用 Intellij IDEA 进行 Scala 开发之前，需要查检 Scala 插件是否正常。启动 Intellij IDEA，然后单击"File->Setting"，选择"plugins"，输入 Scala 可以查看 Scala 插件是否已经安装，如图 1-13 所示。如果你的 Intellij IDEA 版本没有自带 Scala 插件，则可以在此进行安装。

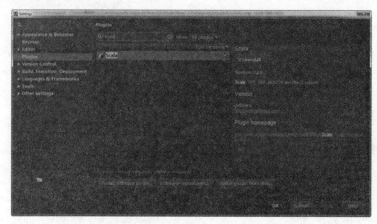

图 1-13　查看 Intellij IDEA 15.0.4 中的 Scala 插件

1.3　Scala Hello World

1.3.1　创建 Scala Project

选择 File | New | Project，就可以看到 New Project 界面，如图 1-14 所示，在左侧工程类型栏选择"Scala"，右边也选择"Scala"，

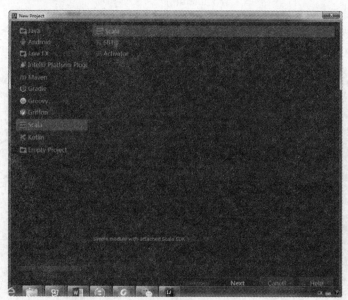

图 1-14　New Project 界面

然后单击"Next"按钮，进入工程配置页面，如图 1-15 所示。

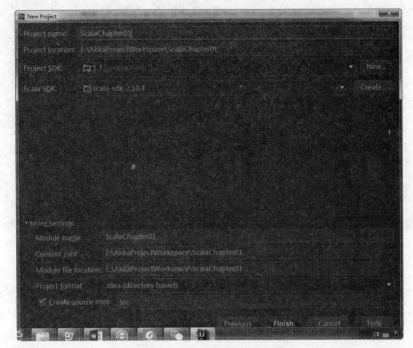

图 1-15　Scala 工程配置页面

选择项目存放位置 Project Location，输入项目名称，指定前面 JDK 安装目录设置 Project SDK，指定前面 Scala SDK 安装目录设置 Scala SDK，然后单击 Finish 按钮完成项目的创建，如图 1-15 所示。

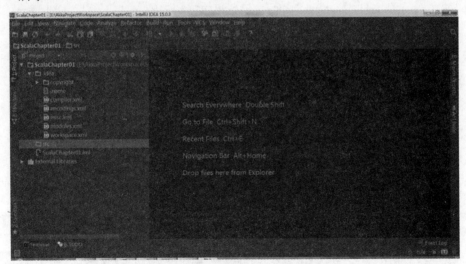

图 1-16　ScalaChapter01 项目结构

1.3.2　配置项目代码目录结构

通过 File|Project Structre 打开项目配置页面，如图 1-17 所示。

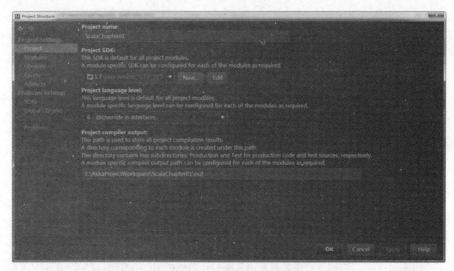

图 1-17　Project Struture 页面

在 Projet Settings 中有 Project、Modules、Libraries、Facets 及 Artifacts 5 项内容。Project 用于设置 Project SDK、编译输出文件目录等；Modules 用于将大型工程分成多个模块，模块相互之间可能有依赖关系，各模块有自己的代码、单元测试代码、资源及脚本等；Libraries 用于配置项目的依赖包；Facets 用于提供特定框架如 Web、Spring、Persistence 的代码帮助，一般由 IDE 自动配置；Artifacts 为工程生成的输出文件，可以是 Java archive（JAR）、Web Application Archive（WAR）及 Enterprise Archive（EAR）等，如果实际中使用了 Maven 等构建工具，则不需要手动配置，否则需要在此手动配置输出。

在 Modules 中，将项目代码目录结构配置为图 1-18 所示，源代码目录与测试代码分开放，Java 代码放在 java 源文件中，Scala 代码放在 Scala 源文件中。

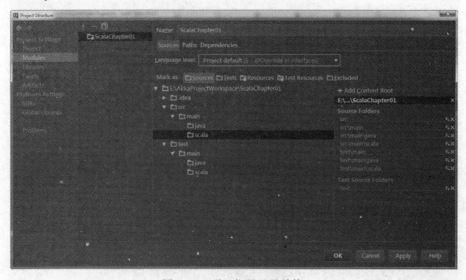

图 1-18　项目代码目录结构

1.3.3 创建应用程序对象

单击 src/main/scala 文件夹，右键|New|Scala Class，如图 1-19 所示。

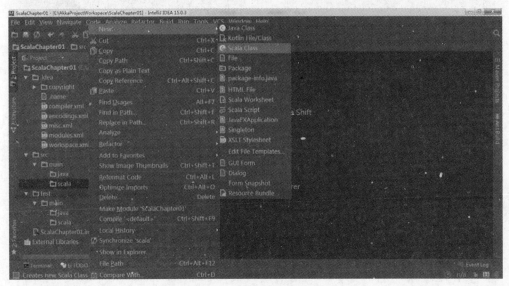

图 1-19　创建 Scala Class

打开 Create New Scala Class 对话框，如图 1-20 所示，在 Name 中输入 HelloWorld，在 Kind 中选择 Object。

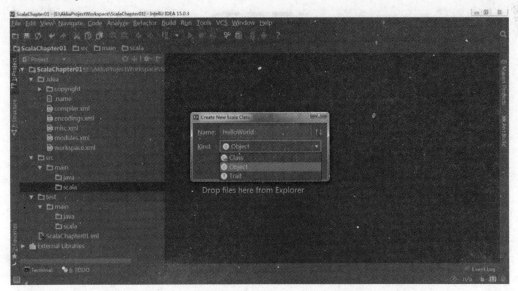

图 1-20　Create New Scala Class 对话框

完成后单击"OK"，这样便创建好了 object HelloWorld，如图 1-21 所示。

图 1-21 object HelloWorld

在编辑框内输入下列代码：

```
def main(args: Array[String]) {
    println("Hello Scala World!")
}
```

完整代码如图 1-22 所示。

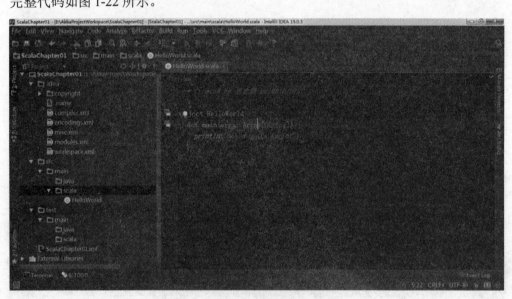

图 1-22 Scala HelloWorld 完整代码

1.3.4 运行代码

通过快捷键 Ctrl+Shift+F10 或在代码编辑器中右键选择 Run 'HelloWorld'，如图 1-23 所示。

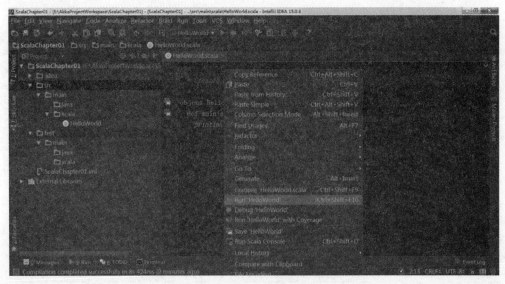

图 1-23　运行 HelloWorld 代码

代码会先经过编译然后运行，运行结果如图 1-24 所示，运行过程中编译生成的字码文件会放置在 out 文件夹中。

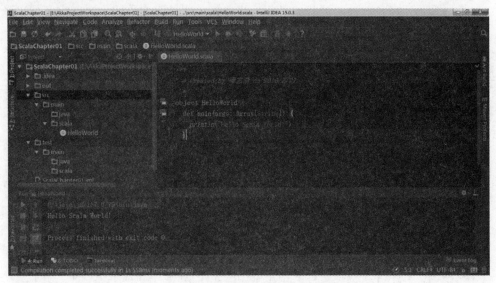

图 1-24　HelloWorld 程序运行结果

1.4　Intellij IDEA 常用快捷键

1.4.1　代码编辑类常用快捷键

表 1-2 给出的是代码编辑器中的常用快捷键，其中最为常用的有：Ctrl+/、Ctrl+Shift+/、Ctrl+F、Ctrl+R、Alt+Enter。

表 1-2 代码编辑类常用快捷键

快捷键	作用
Ctrl+Shift+Up（上移） Ctrl+Shift+Down（下移）	移动选中的代码
Ctrl+D	复制选中的代码
Ctrl+Y	删除选中的代码
Ctrl+/	将选中的代码作为注释（未注释时）；删除选中代码注释（已注释时）。适用于//注释语法
Ctrl+Shift+/	将选中的代码作为注释（未注释时）；删除选中代码注释（已注释时）。适用于/**/注释语法
Ctrl+F	在当前代码文件中查找
Ctrl+R	在当前代码文件中查找并替换
Alt+左箭头	切换到前一 Tab 页打开的文件
Alt+右箭头	切换到后一 Tab 页打开的文件
Ctrl+Alt+T	使用 if 语句、for 循环语句、while 循环语句、try catch 等语句包裹当前选中的代码
Alt+Enter	代码快速修正

Ctrl+/：使用//注释符对选中的代码进行注释或删除注释，如果代码之前没有被注释，则会加上注释，如果代码已经被注释，则会删除注释。

Ctrl+Shift+/：使用/**/注释符对选中的代码进行注释或删除注释，如果代码之前没有被注释，则会加上注释，如果代码已经被注释，则会删除注释。

Ctrl+F：用于在当前代码文件中查找，查找时可以选择是否区别大小写等，还可以使用正则表达式进行查找。

Ctrl+R：用于在当前代码文件中查找并替换，同样也可以指定大小写，是否启用正则表达式，替换时可以选择逐个替换或全部替换。

Alt+Enter：用于在代码出错时快速地给出修正提示，如使用的类没有引入时会弹出提示选项，供开发人员选择。

1.4.2 导航快捷键

表 1-3 给出了常用导航快捷键，其中最常用的是 Ctrl+N、Ctrl+H。Ctrl+N 用于快速定位到相关类所在的代码文件，Ctrl+H 则用于显示类的继承关系结构。

表 1-3 导航常用快捷键

快捷键	作用
Ctrl+E	显示最近访问的文件
Ctrl+N	导航到指定的类
Ctrl+Shift+N	导航到指定的文件或文件夹
Ctrl+Shift+Alt+N	导航到指定的方法或成员变量
Ctrl+H	显示类的继承关系
Shift+Shift	在整个工程中包括资源文件、库文件等中查找

1.4.3 编译、运行及调试

表 1-4 中给出了在编译、运行及调试时常用的快捷键。

表 1-4 编译、运行及调试常用快捷键

快捷键	作用
Ctrl+F9	编译整个工程项目
Shift+F10	运行当前代码文件
Shift+F9	调试当前代码文件
Ctrl+F8	设置断点
F7	Step into
F8	Step out
F9	恢复执行
Alt+F8	表达式计算（Evaluate expression）

1.4.4 代码格式化

代码格式化常用快捷键，如表 1-5 所示。

表 1-5 代码格式化常用快捷键

快捷键	作用
Ctrl+Alt+L	格式化代码
Ctrl+Alt+I	代码自动缩进
Ctrl+Alt+O	引入优化（例如删除未使用的引入）

1.5 交互式命令行使用

与 Java 语言相比，Scala 提供了一个非常重要的语言学习工具，那就是 REPL 命令行（Read-Evaluate-Print-Loop Command），也称交互式命令行，它有点类似于 Linux 命令行，执行完程序后会立即显示程序运行结果，该工具对 Scala 语言学习者了解 Scala 语言特性大有裨益。本节将对 3 种常见的 Scala 语言交互式命令行进行介绍。

1.5.1 Scala 内置交互式命令行

在 Windows 系统上，可以通过 DOS 命令行进入 Scala SDK 提供的 Scala 交互式命令行，直接在命令行上输入 scala，然后回车便能进入 Scala 交互式命令行，如图 1-25、图 1-26 所示。

图 1-25 进入 Scala 交互式命令行界面

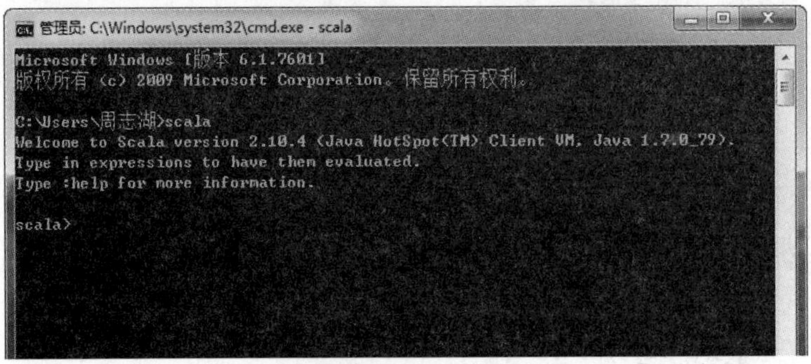

图 1-26 进入 REPL 命令行

进入命令行之后，可以直接在命令行上输入要执行的命令，然后按回车键，便能看到程序的执行结果，如图 1-27 所示。

图 1-27 REPL 命令行程序运行结果

在 REPL 命令行中可以通过:help 命令来获取帮助。

```
scala> :help
All commands can be abbreviated, e.g. :he instead of :help.
Those marked with a * have more detailed help, e.g. :help imports.

:cp <path>                add a jar or directory to the classpath
:help [command]           print this summary or command-specific help
:history [num]            show the history (optional num is commands to show)
:h? <string>              search the history
```

```
:imports [name name ...]    show import history, identifying sources of names
:implicits [-v]             show the implicits in scope
:javap <path|class>         disassemble a file or class name
:load <path>                load and interpret a Scala file
:paste                      enter paste mode: all input up to ctrl-D compiled tog
ether
:power                      enable power user mode
:quit                       exit the interpreter
:replay                     reset execution and replay all previous commands
:reset                      reset the repl to its initial state, forgetting all s
ession entries
:sh <command line>          run a shell command (result is implicitly => List
[Str
ing])
:silent                     disable/enable automatic printing of results
:type [-v] <expr>           display the type of an expression without evaluating
 it
:warnings                   show the suppressed warnings from the most recent lin
e which had any
```

1.5.2 Scala Console 交互式命令行

Scala SDK 提供的 REPL 命令行虽然很强大，但美中不足的是，当编写的代码较复杂的时侯，不能够像在 IDE 里面那样给出相应的代码提示。为解决这一问题，Intellij IDEA 中提供了 Scala Console 交互式命令行。可以使用快捷键 Ctrl+Shift+D 或通过 Tools|Run Scala Console（如图 1-28 所示）来运行 Scala Console。

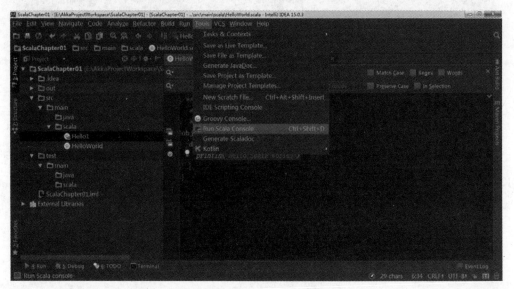

图 1-28　Run Scala Console

运行成功后如图 1-29 所示。

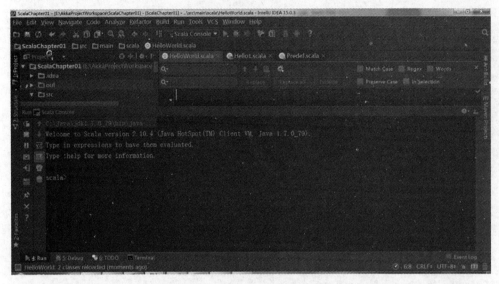

图 1-29　Scala Console 运行界面

Scala Console 最大的特点是在输入代码时，会给出相应的提示，如图 1-30 所示。

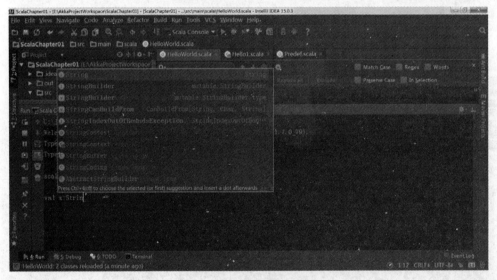

图 1-30　Scala Console 中的代码自动提示

代码输入完成后，按 Ctrl+Enter 键便会执行输入的代码并给出结果，如图 1-31 所示。

Scala Console 还有一个特点，就是可以回车换行，编写多个程序语句一起执行，而在 Scala SDK 提供的交互式命令行中如果要编写多个语句一起执行的话，需要使用;号将各条语句隔开或使用:paste 命令。

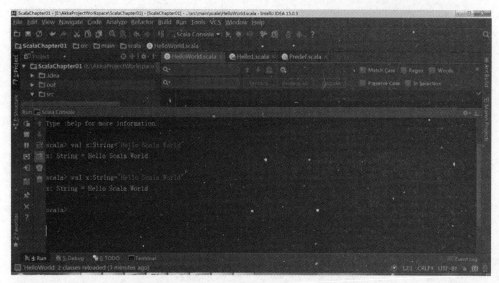

图 1-31　Scala Console 中的代码运行结果

1.5.3　Scala Worksheet 的使用

在学习 Scala 时，还有一种比较方便的工具能够辅助我们快速得到 Scala 程序的运行结果，这便是 Scala Worksheet，它通过 File|New|Scala Worksheet 创建，如图 1-32 所示，然后输入 Worksheet 的名称（本例中命名为 HelloWorldWorksheet）。

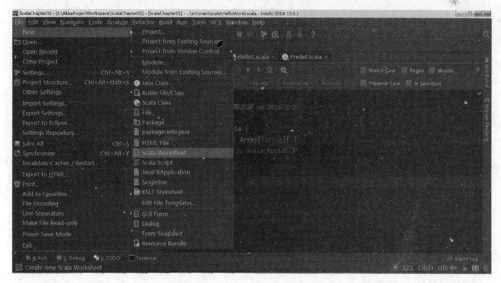

图 1-32　创建 Scala Worksheet

创建完成后，在编辑器中输入 Scala 代码，然后保存，这样就能够在右侧看到代码运行结果，如图 1-33 所示。

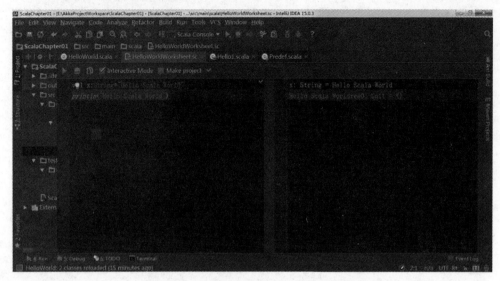

图 1-33　Scala Worksheet 中代码运行结果

在这 3 种交互式命令行中，Scala SDK 提供的 REPL 命令行无代码提示功能而且执行多行代码时不方便，Scala WorkSheet 需要创建 Worksheet 文件，运行完成后代码能够永久保留但运行速度较慢，特别是代码量较大时，每保存一次都要重新执行文件中的所有代码，而 Scala Console 在代码提示、代码执行速度方面都有明显的优势，因此建议大家使用 Intellij IDEA 提供的 Scala Console 交互式命令行来学习 Scala 语言。

小　　结

本章对 Scala 语言及其特点进行了介绍，给出了如何在 Windows 环境下进行 Scala 开发环境的搭建，然后演示了 Scala 版本 HelloWorld 程序的编程与运行，对集成开发环境 Intellij IDEA 中的常用快捷键进行了介绍，最后对 Scala 中原生的交互式命令行、Intelli IDEA 中提供的 Scala Console 交互式命令行及 Scala Worksheet 的使用进行了说明。通过本章的学习，读者可以简单了解 Scala 语言的特点，掌握 Scala 开发环境的搭建及 Scala 常用交互式命令行的使用。

第 2 章 变量及基本数据类型

本章将介绍 Scala 变量的定义，Scala 中的基本数据类型及对应的基本操作，Scala 中的特殊变量类型和运算符的优先级。

2.1 变量

2.1.1 变量定义

Scala 中有两种类型的变量：可变变量和不可变变量。可变变量指的是变量被赋值后，变量值可以随着程序的运行而改变，可变变量使用关键字 var 进行定义；不可变变量指的是变量一旦被赋值，变量值在程序运行的过程中不会被改变，不可变变量使用关键字 val 进行定义。

下面示例为使用关键字 val 定义不可变变量的情况。

```
//声明一个 val 变量，Scala 会自动进行类型推断
scala> val helloString="Hello World"
helloString: String = Hello World

//声明一个 val 变量，明确指定变量类型
scala> val helloString:String="Hello World"
helloString: String = Hello World

//String 其实就是 java.lang.String，Scala 会默认引入 java.lang 包
scala> val helloString:java.lang.String="Hello World"
helloString: String = Hello World

//不能被重新赋值，因为它是 val 变量
scala> helloString="Hello Crazy World"
<console>:8: error: reassignment to val
       helloString="Hello Crazy World"
```

通过上述代码可以看到，Scala 中的 val 变量与 Java 语言中 final 关键字修饰的变量一样，变量一旦被定义赋值，其值在程序运行过程中便不能被改变。如果在程序中使用的变量随程序运行过程而改变，则可以使用 var 关键字定义变量。例如：

```
//var 声明可变变量
scala> var helloString="Hello Cruel World"
helloString: String = Hello Cruel World

//重新赋值
scala> helloString="GoodBye Cruel World"
helloString: String = GoodBye Cruel World
```

通过上述代码可以看到，Scala 中使用 var 关键字定义的变量在程序运行过程中值可以随程序的运行而发生改变。

需要特别注意的是，Scala 中的变量在定义时必须初始化赋值。

```
//变量必须初始化赋值
scala> val s2:String
```

```
<console>:7: error: only classes can have declared but undefined members
       val s2:String

scala> val s2:String=""
s2: String = ""

//对 var 类型的变量同样适用
scala> var s2:String
<console>:7: error: only classes can have declared but undefined members
(Note that variables need to be initialized to be defined)
       var s2:String
```

另外，变量定义初始化还存在一种比较特殊的方式，那就是使用占位符"_"。

```
//String 类型的变量使用占位符初始化
scala> var s2:String=_
s2: String = null

//Int 类型的变量使用占位符初始化
scala> var i1:Int=_
i1: Int = 0

//Float 类型的变量使用占位符初始化
scala> var f1:Float=_
f1: Float = 0.0

scala> var c1:Char=_
c1: Char = ?
```

通过上面的代码可以看到，基本类型 String 使用"_"初始化时变量为 null，其他如 Float、Int、Double 等基本数据类型则被初始化为 0，Char 类型被初始化为?。

2.1.2　lazy 变量

Scala 中的变量还可以使用 lazy 关键字来修饰，经过 lazy 关键字修饰的变量只有在真正使用时才会被赋值，具体见下列代码。

```
//普通 val 变量定义
scala> val v1="test"
v1: String = test

//经过 lazy 关键字修饰的变量在定义时不被赋值
scala> lazy val v1="test"
v1: String = <lazy>

//在使用时，变量才会赋值
```

```
scala> v1
res0: String = test
```

代码 lazy val v1="test" 定义了一个 val 类型的变量,该变量在定义时没有赋值,其返回结果为 v1: String = <lazy>,在使用时才被赋值。可以看到,与普通变量所不同的是,普通变量赋值后立即会得到赋值结果。值得注意的是 lazy 关键字只能修饰 val 类型的变量,而不能修饰 var 类型的变量。例如:

```
scala> lazy var v2="test2"
<console>:1: error: lazy not allowed here. Only vals can be lazy
       lazy var v2="test2"
```

lazy 关键字不能用于 var 类型变量主要是为避免程序运行过程中变量未使用便被重新赋值。

2.2 基本数据类型

Scala 提供的基本数据类型如表 2-1 所示。

表 2-1 Scala 基本数据类型

数据类型	取值范围
Byte 字节型	8 位有符号的整数($-2^7 \sim 2^7-1$)
Short 短整型	16 位有符号的整数($-2^{15} \sim 2^{15}-1$)
Int 整型	32 位有符号的整数($-2^{31} \sim 2^{31}-1$)
Long 长整型	64 位有符号的整数($-2^{63} \sim 2^{63}-1$)
Char 字符型	16 位无符号字符数($0 \sim 2^{16}-1$)
String 字符串型	连续字符串
Float 浮点型	32 位浮点数
Double 双精度浮点型	64 位浮点数
Boolean 布尔型	真(true),假(false)

从表中可以看出,Scala 的基本数据类型与 Java 中的基本数据类型是一一对应的,不同的是 Java 的数据类型首字母不需要大写,而 Scala 的首字母必须大写,这是因为 Scala 中所有的值类型都是对象。

本节以 Int、Float、Double、Char、String 及 Boolean 为例介绍 Scala 的基本数据类型及变量定义。

2.2.1 Int 类型

Int 类型对应的变量为整型变量,变量对应的是整型数据。Int 类型变量的定义有多种方式,包括十六进制、十进制、八进制定义法。

下面的代码给出的是整型变量的十六进制定义法。

```
//16 进制定义法
```

```
scala> val x=0x29
x: Int = 41
```

下面的代码给出的是十进制定义法。

```
//10 进制定义法
scala> val x=41
x: Int = 41
```

下面的代码给出的是八进制定义法。

```
//8 进制定义法
scala> 051
res0: Int = 41
```

在实际中使用最多的整型变量定义方式当然是十进制定义法。

2.2.2 Float 类型

Float 类型表示的是浮点数。如果直接输入浮点数，则 Scala 编译器会自动进行类型推导，并自动解释成 Double 类型，需要在浮点数后加 F 或 f 才能定义 Float 类型的变量。Float 类型变量定义如下：

```
//要定义 Float 类型浮点数，需要在浮点数后面加 F 或 f
scala> val floatNumber=3.141529F
floatNumber: Float = 3.141529


//小写的 f 也可以
scala> val floatNumber=3.141529f
floatNumber: Float = 3.141529
```

2.2.3 Double 类型

Double 数据类型表示的是双精度的浮点数。Double 类型变量的定义如下所示：

```
//Double 类型定义,直接输入浮点数，编译器会将其自动推断为 Double 类型
scala> val doubleNumber=3.141529
doubleNumber: Double = 3.141529
```

双精度浮点类型的变量还可以采用指数表示法，浮点数后加 E 或 e 均可，例如：

```
//浮点数指数表示法，e 也可以是大写 E,0.314529e1 与 0.314529*10 等同
scala> val floatNumber=0.314529e1
floatNumber: Double = 3.14529
```

2.2.4 Char 类型

Char 数据类型表示的是字符类型，用单引号将字符包裹起来。Char 类型变量定义如下所示：

```
//字符定义,用''将字符包裹
scala> var charLiteral='A'
charLiteral: Char = A
```

部分特殊字符如双引号、换行符及反斜杠等的定义需要加转义符\或使用对应的 Unicode 编码,下面的示例给出了双引号字符的定义:

```
//通过转义符\进行双引号的定义
scala> var x='\"'
x: Char = "


//通过使用Unicode编码进行双引号的定义
scala> var y='\u0022'
y: Char = "
```

> **提示** 常用特殊字符包括:
> \n 换行符,其 Unicode 编码为 (\u000A)
> \b 回退符,其 Unicode 编码为 (\u0008)
> \t tab 制表符,其 Unicode 编码(\u0009)
> \" 双引号,其 Unicode 编码为 (\u0022)
> \' 单引号,其 Unicode 编码为 (\u0027)
> \ 反斜杠,其 Unicode 编码为(\u005C)

2.2.5 String 类型

String 数据类型表示的是字符串类型,用双引号" "将字符串包裹起来。String 类型变量的定义如下所示:

```
//字符串变量用""包裹
scala> val helloWorld="Hello World"
helloWorld: String = Hello World
```

如果字符串类型中有双引号,则需要使用转义符\。

```
//要定义"Hello World",可以加入转义符\
scala> val helloWorldDoubleQuote="\"Hello World\""
helloWorldDoubleQuote: String = "Hello World"
```

Scala 中还提供了一种原样输出字符串内容的语法,用 3 个双引号""""将字符串包裹起来。

```
//如果希望能够原样输出字符串中的内容,则用三个引号"""将字符串包裹起来,如
scala> println(""" hello cruel world, \n \\\\ \b \\, I am " experienced" programmer""")
hello cruel world, \n \\\\ \b \\, I am " experienced" programmer
```

2.2.6 Boolean 类型

Boolean 数据类型表示的是布尔类型，Boolean 类型变量的定义如下所示：

```
scala> var x=true
x: Boolean = true
```

2.3 基本类型操作

Scala 语言是纯面向对象编程语言，在 Scala 中一切皆为对象，所有的操作都是方法调用。本节将对基本数据类型的操作进行介绍，包括算术运算操作、关系运算操作、逻辑运算操作、位运算操作、对象比较运算操作及字符串比较运算操作。

2.3.1 算术运算操作

下面给出的代码是基本的算术运算操作，包括加（+）、减（-）、乘（*）、除（/）、取模（%）。

```
//整数求和，编译器会将其转换为(1).+(2)执行，类似Java的方法调用。
scala> var sumVlaue=1+2
sumVlaue: Int = 3

//前一语句等同于下列语句
scala> var sumVlaue=(1).+(2)
sumVlaue: Int = 3

//操作符重载,编译器会将其转换为(1).+(2L)执行
scala> val longSum = 1 + 2L
longSum: Long = 3

//减法
scala> 1-3
res5: Int = -2

//除法
scala> 1/3
res6: Int = 0

//取模
scala> 1%3
res7: Int = 1
```

```
//乘法
scala> 1L*3L
res8: Long = 3
```

Scala 语言还有一个非常值得注意的地方,那就是它提供了用+、- 符号来表示正负数,并且这两个符号可以直接在操作中使用,例如:

```
//scala 中可以用+、-符号来表示正负数,例如-3、+3,并且可以加入到运算符当中
//1+ -3 编译器将-3 解释成一个负数。
scala> var y=1+ -3
y: Int = -2
```

2.3.2 关系运算操作

Scala 的关系运算操作包括大于(>)、小于(<)、小于等于(<=)、大于等于(>=),具体使用如下所示:

```
//>运算符
scala> 3 > -3
res12: Boolean = true

//<运算符
scala> 3 < -3
res13: Boolean = false

//<=运算符
scala> 3 <= -3
res14: Boolean = false

//<=运算符
scala> 3 <=3
res15: Boolean = true

//<=运算符,! 为取反操作
scala> !(3<= -3)
res16: Boolean = false
```

2.3.3 逻辑运算操作

逻辑运算操作包括逻辑与(&&)及逻辑或(||)运算符,逻辑与操作为真时必须保证两个变量同时为真,而逻辑或操作为真时只要求至少有一个变量为真,具体使用如下所示:

```
scala> val bool=true
bool: Boolean = true
```

```
//逻辑与&&: 同时为true时才会true
scala> bool && bool
res17: Boolean = true

scala> bool && !bool
re:18: Boolean = false

//逻辑或||: 同时为flase时才为false
scala> bool || bool
res19: Boolean = true

scala> bool || !bool
res20: Boolean = true

scala> false || false
res21: Boolean = false
```

2.3.4 位运算操作

位运算操作包括位与（&）、位或（|）、位异或（^）、取反（~）、左移位（<<）、右移位（>>）、无符号左移位（<<<）及无符号右移位（>>>）操作，具体示例如下所示：

```
//整型1对应的二进制:      00000000 00000000 00000000 00000011
//整形2对应的二进制:      00000000 00000000 00000000 00000101
//位与（&）操作后的二进制: 00000000 00000000 00000000 00000001
scala> 3&5
res25: Int = 1

//整型1对应的二进制:      00000000 00000000 00000000 00000011
//整形2对应的二进制:      00000000 00000000 00000000 00000101
//位或（|）操作后的二进制: 00000000 00000000 00000000 00000111
scala> 3 | 5
res26: Int = 7

//整型1对应的二进制:       00000000 00000000 00000000 00000011
//整形2对应的二进制:       00000000 00000000 00000000 00000101
//位异或（^）操作后的二进制: 00000000 00000000 00000000 00000110
scala> 3^5
res27: Int = 6

//整型3对应的二进制:  00000000 00000000 00000000 00000011
//取反后的二进制:     11111111 11111111 11111111 11111100
scala> ~3
```

```
res28: Int = -4

//左移位（shift left）
//00000000 00000000 00000000 00000110
//00000000 00000000 00000000 00001100
scala> 6 << 1
res29: Int = 12

//右移位（shift left）
//00000000 00000000 00000000 00000110
//00000000 00000000 00000000 00000011
scala> 6 >> 1
res28: Int = 3

//无符号右移（shift left）
//11111111 11111111 11111111 11111111
//00000000 00000000 00000000 00000001
scala> -1 >>> 31
res32: Int = 1
```

2.3.5 对象比较运算操作

Scala 中的对象比较不同于 Java 中的对象比较，Scala 是基于内容比较，而 Java 中比较的是引用，即对象的物理内存地址是否一样，进行内容比较时须定义比较方法。基于内容的比较是使用 Scala 语言特别需要注意的，它是 Scala 语言的重要特点之一。

```
scala> 1==1
res34: Boolean = true

scala> 1==1.0
res35: Boolean = true

scala> val x="Hello"
x: String = Hello

scala> val y="Hello"
y: String = Hello

//Scala 基于内容比较，而 Java 中比较的是引用，进行内容比较时须定义比较方法
scala> x==y
res36: Boolean = true

scala> var x=new String("Hello")
x: String = Hello
```

```
scala> var y=new String("Hello")
y: String = Hello

//equals 方法等同于==
scala> y.equals(x)
res37: Boolean = true

scala> y==x
res38: Boolean = true

//如果需要像 Java 一样，比较内存地址即引用的是否为同一个对象的话，则使用 eq 方法
scala> y.eq(x)
res39: Boolean = false
```

2.3.6 字符串运算操作

Scala 中定义的 String 类型实际上其实现就是 java.lang.String 类型，因此可以调用 Java 中 String 类型所有的方法。例如：

```
scala> var str="Hello"
str: String = Hello

scala> str.indexOf("o")
res12: Int = 4

scala> str toUpperCase
warning: there were 1 feature warning(s); re-run with -feature for details
res53: String = HELLO
```

也可以调用以下方法：

```
//反转符
scala> res53.toLowerCase
res54: String = hello

//换位符
scala> str.reverse
res55: String = olleH

//丢弃字符
scala> str.drop 3
res57: String = lo

//获取一定范围内的子串
```

```
scala> str slice<1,4>
res58: String = ell
```

熟悉 Java 语言的读者可能清楚，Java.lang.String 中并没有定义 reverse、drop 等这些方法，Scala 是如何实现的呢？事实上，Scala 会将 String 类型对象转换成 StringOps 类型对象，在遇到 reverse、map、drop 和 slice 等方法调用时编译器会自动进行隐式转换。关于隐式转换，在后续章节中会详细介绍。

2.4 运算符的优先级

在 Scala 中运算符优先级如表 2-2 所示，其中，*、/、%优先级最高，以此类推，部分运算符如::（List 构造）、:::（List 拼接）在第 4 章讲解集合会有涉及，这里只要知道其运算优先级即可。

表 2-2 Scala 运算符优先级

优先级（从高到低）	符号及描述
1	*（乘）、/（除）、%（取模）
2	+（加）、-（减）
3	::（List 构造）、:::（List 拼接），它们是右操作，例如 a::b::Nil，执行顺序为(a::(b::Nil))
4	<=（小于等于）、>=（大于等于）、==（判断是否相等）、=（赋值）、!（取反）、!=（判断是否不等）
5	<（小于）、>（大于）、<<（左移位）、>>（右移位）、<<<（无符左移位）、>>>（无符右移位）
6	&（位与）、&&（逻辑与）
7	^（位异或）
8	|（位或）、||（逻辑或）

在执行程序时，如果存在多个同一优先级的运算，则按照从左到右的执行顺序（:::操作是个例外，它是从右到左的执行顺序），具体代码如下：

```
//从左到右执行程序，先进行取模运算%，再进行乘积*
scala> 1 % 2 * 3
res0: Int = 3
```

如果想改变运算的优先级，可以在程序中使用()。

```
//使用()改变运算的优先级，先执行 2*3，再进行取模运算%
scala> 1 % (2 * 3)
res1: Int = 1
```

再看一个复杂的例子，代码如下：

```
scala> 1*10-2::2+2::Nil:::List(2,3):::List(5,6)
```

```
res13: List[Int] = List(8, 4, 2, 3, 5, 6)
```

按前述的优先级顺序，它首先执行的是 1*10，得到 10-2::2+2::Nil:::List(2,3):::List(5,6)，然后执行 10-2，得到 8::2+2::Nil:::List(2,3):::List(5,6)，再执行 2+2，得到 8::4::Nil:::List(2,3):::List(5,6)，然后执行 8::4::Nil，根据右优先执行顺序，实际执行顺序为 (8::(4::Nil))，得到 List(8,4) :::List(2,3):::List(5,6)，然后按照右优先顺序，实际执行顺序为 (List(8,4) :::(List(2,3):::List(5,6)))，最终得到 List(8, 4, 2, 3, 5, 6)。

2.5 元组类型

Scala 中还有一种非常特殊的类型，称为元组。元组是不同类型值的聚集，它可以将不同类型的值放在一个变量中保存。元组的定义如下：

```
//元组的定义
scala> ("hello","china","beijing")
res23: (String, String, String) = (hello,china,beijing)

scala> ("hello","china",1)
res24: (String, String, Int) = (hello,china,1)

scala> var tuple=("Hello","China",1)
tuple: (String, String, Int) = (Hello,China,1)
```

要访问元组的内容，可以通过变量名._N 的方式进行，其中 N 表示元组中元素的索引号，例如：

```
//访问元组内容
scala> tuple._1
res25: String = Hello

scala> tuple._2
res26: String = China

scala> tuple._3
res27: Int = 1
```

在使用时还可以将元组的内容进行提取，对变量进行初始化，代码如下：

```
//通过模式匹配获取元组内容
scala> val (first, second, third)=tuple
first: String = Hello
second: String = China
third: Int = 1
```

代码 val (first, second, third)=tuple 通过元组对变量进行赋值，从而提取元组中的所有元素，

(first, second, third)中的变量分别对应元组中的第一个元素、第二个元素及第三个元素。对于元组类型，使用==进行比较的时候，进行的是内容比较，这点与 String 类型的变量比较是一致的，例如：

```
scala> var tuple=("Hello","China",1)
tuple: (String, String, Int) = (Hello,China,1)

scala> var tuple1=("Hello","China",1)
tuple1: (String, String, Int) = (Hello,China,1)

scala> tuple==tuple1
res4: Boolean = true
```

2.6 符号类型

Scala 语言中还存在一种常用类型，即符号（Symbol）类型，符号类型的定义需要使用'符号，例如：

```
//使用'定义符号类型的变量，Scala 类型推断为 Symbol 类型
scala> val s='start
s: Symbol = 'start

//明确指定为 Symbol 类型
scala> val s1:Symbol='stop
s1: Symbol = 'stop
```

符号类型主要起标识作用，在模式匹配、内容判断中比较常用，例如：

```
scala> if(s1=='start) println("Start......") else println("other......")
Start......
```

符号类型变量在输出时，会原样输出，例如：

```
//符号类型变量会原样输出
scala> println(s1)
'start
```

另外与 String 类型、元组类型等变量一样，使用==符号进行变量比较时，比较的是变量的内容而非引用。

```
scala> val s2='Start
s2: Symbol = 'Start

scala> val s3='Start
```

```
s3: Symbol = 'Start

//==比较的是符号变量的内容而非引用
scala> s2==s3
res6: Boolean = true
```

小　结

通过本章的学习，读者应该掌握 val、var 类型变量的定义与使用，掌握 Scala 中的基本数据类型及它们与 Java 语言中的基本类型变量的对应关系，学会使用 Scala 基本数据类型对应的操作，熟悉常用数学运算符号的优先级，学会使用元组类型、符号类型。在下一章中，我们将学习 Scala 语言中的程序控制结构。

第 3 章 程序控制结构

本章介绍程序的基本控制结构。包括：If 语句、while 语句、do while 语句、for 循环语句的使用。

3.1　if 语句

If 语句在实际编程中应用十分广泛，是构成程序逻辑的基础，下面就 Scala 中的 if 语句、if…else…语句、if…else if…else…语句、if 嵌套语句及 if 语句作为表达式的用法进行详细介绍。

3.1.1　if 语句

语法格式如下：

```
if(条件判断)
{
    //条件判断为真时执行
}
```

示例如下：

```
scala> val x=8
x: Int = 8

scala> if(x < 10) println(s"$x is smaller than 10")
8 is smaller than 10
```

3.1.2　if…else…语句

语法格式如下：

```
if(条件判断){
    //条件判断为真时执行
}else{
    //条件判断为假时执行
}
```

示例如下：

```
scala> val x=9
x: Int = 9

scala> if(x<8) println("small") else  println("big")
big
```

3.1.3　if...else if...else…语句

语法格式如下：

```
if(条件判断语句1){
```

```
    //条件判断语句 1 为真时执行
}else if(条件判断语句 2){
    //条件判断语句 2 为真时执行
}else if(条件判断语句 3){
    //条件判断语句 2 为真时执行
}else {
    //前面的条件判断都为假时执行
}
```

示例如下：

```
scala> if(x==8)
  println("8")
else if (x==7)
  println("7")
else if (x==9)
  println("9")
else
  println("other")

9
```

3.1.4　if 的嵌套使用

语法格式如下：

```
if(条件判断 1){
    if(条件判断 2){
        //条件判断 2 为真时执行
    }else{
        //条件判断 2 为假时执行
    }
}else{
    //条件判断 1 为假时执行
}
```

示例如下：

```
val x=9
if(x<10)
    if(x==9)
      println("9")
    else
      println("other")
 else
    println("bigger than 9")
```

3.1.5 if 表达式

与 Java、C++、C 等高级程序设计语言的 if 条件判断语句所不同的是，Scala 中的 if 语句可以作为表达式使用，表达式具有返回值可以直接赋值给变量的功能，例如：

```
scala> val x= if("hello"=="hell")  1 else 0
x: Int = 0

scala> val x= if("hello"=="hell")  println("1") else println("0")
0
x: Unit = ()
```

通过上述示例可以看到，if 是个表达式，其返回值可以给变量赋值。Scala 会将 if 语句最后一条执行语句作为返回值。

3.2 while 循环语句

Scala 语言也提供了 while 循环语句，不过 Scala 中弱化了 while 循环语句的作用，在程序中不推荐使用 while 循环，尽量使用 for 循环或递归来替代 while 循环语句。

3.2.1 while 语句的使用

语法格式如下：

```
while(条件判断){
    //条件判断为真时执行
}
```

示例代码：

```
var i=15
    while(i<20){
        println("i="+i)
        i=i+1
    }
```

运行结果：

```
i=15
i=16
i=17
i=18
i=19
```

3.2.2　do while 语句的使用

语法格式如下:

```
do{
    //先执行，再进行条件判断，如果为真则继续循环执行
}while(条件判断)
```

示例代码:

```
var j=15
    do{
      println("i="+i)
      i=i+1
    }while(i<20)
```

运行结果:

```
i=15
i=16
i=17
i=18
i=19
i=20
```

Scala 中 while 和 do while 的语法与 Java、C++是一样的，while 先判断后执行，do while 先执行后判断。

值得注意的是，与 if 不同，while 与 do while 也有返回值，只不过其返回值始终为 Unit。

```
scala> var i=15
i: Int = 15

//while 循环有返回值，其返回值始终为 Unit
scala> var x=while(i<20){
  println("i="+i)
  i=i+1
}
i=15
i=16
i=17
i=18
i=19
x: Unit = ()
```

在某些纯函数式编程语言中，删除了 while 与 do while 程序控制结构，但 Scala 仍然保留了 while 与 do while，我们知道函数式编程语言推崇使用 val 类型的变量，而 while 循环语句

需要和 var 类型的变量一起使用，可见 Scala 并不是纯函数式编程语言。

3.3　for 循环语句

3.3.1　基础 for 循环

Scala 中没有 C、C++及 Java 等高级语言中的 for（初始化变量;条件判断;更新变量）循环，而是有自己独特的 for 循环风格，Scala 中的基础 for 循环语法格式如下：

```
for (i <- 表达式){
    //执行循环中的语句
}
```

示例代码：

```
    for(i<- 1 to 5){
      println("i="+i)
    }
```

运行结果：

```
i=1
i=2
i=3
i=4
i=5
```

程序中的<-被称为生成器（generator），for 循环实际上是通过对集合的遍历来达到循环的目的，它首先会执行 1 to 5，相当于调用 1.to(5)，但我们知道，整型没有 to 方法，此时它会将 Int 类型隐式转换成 scala.runtime.RichInt 类型，然后调用 RichInt 中的 to 方法：

```
def to(end: Int): Range.Inclusive = Range.inclusive(self, end)
```

生成 scala.collection.immutable.Range.Inclusive 集合。在循环时将集合中的各个元素赋值给变量 i，从而完成整个循环。for(i<- 1 to 5)相当于下面这行代码：

```
scala> for(i<- 1.to(5)) println("i="+i)
i=1
i=2
i=3
i=4
i=5
```

也等价于下面两行代码：

```
scala> 1 to 5
res39: scala.collection.immutable.Range.Inclusive = Range(1, 2, 3, 4, 5)
```

```
scala> for(i <- res39) println("Iteration"+i)
Iteration1
Iteration2
Iteration3
Iteration4
Iteration5
```

1 to 5 生成的集合包含了元素 5，即元素范围为[1-5]，如果希望集合是右开的，则可以使用 1 until 5，即元素范围为[1-4]。

```
scala> 1 until 5
res9: scala.collection.immutable.Range = Range(1, 2, 3, 4)

scala> for(i<- res9) println("i="+i)
i=1
i=2
i=3
i=4

scala> for(i<- 1 until 5) println("i="+i)
i=1
i=2
i=3
i=4
```

如果在循环时需要设定步长，则可以使用带步长的方法：

```
scala> for(i<- 1 until(10,2)) println("i="+i)
i=1
i=3
i=5
i=7
i=9
```

代码 1 until(10,2)中的 2 为步长，它调用的是 RichInt 的 until 方法：

```
def until(end: Int, step: Int): Range = Range(self, end, step)
```

同样，对于 RichInt 类型也有相应步长的 to 方法：

```
def to(end: Int, step: Int): Range.Inclusive = Range.inclusive(self, end, step)
```

下面是示例代码：

```
scala> for(i<- 1 to(11,2)) println("i="+i)
i=1
i=3
i=5
```

```
i=7
i=9
i=11
```

无论是 for 循环还是 while 循环,Scala 语言都没有提供 Java 语言中的 break、continue 关键字,如下面的程序:

```
scala> for(i<- 1 to 5){
  println("i="+i)
    if(i!=3) continue;
}
<console>:10: error: not found: value continue
              if(i!=3) continue;

scala> for(i<- 1 to 5){
  println("i="+i)
    if(i!=3) break;
}

<console>:10: error: not found: value break
              if(i!=3) break;
```

在循环中如果需要提供类似 break 语句的功能,可以通过两种途径实现:一是定义 Boolean 类型的变量,在 for 或 while 循环中进行条件判断;二是在程序中引入 scala.util.control.Breaks 类。通过 Boolean 类型的变量退出 for 循环的使用我们会在下一小节中进行介绍,这里先来看如何通过 Breaks 类来达到退出循环。

```
//引入 Breaks 类及所有的方法
  import scala.util.control.Breaks._
  //调用 Breaks 中定义的 breakable 方法
  breakable{
    for(i<- 1 to 5) {
      //break 为 Breaks 中定义的方法
      if(i>2) break
      println("i="+i)
    }
  }
```

程序运行结果如下:

```
i=1
i=2
```

在使用时通过 import scala.util.control.Breaks._ 引入 Breaks 中所有的方法,breakable 为 Breaks 中定义的方法。

```
def breakable(op: => Unit) {
```

```
    try {
      op
    } catch {
      case ex: BreakControl =>
        if (ex ne breakException) throw ex
    }
  }
```

在 for 循环中加入 if(i==3) break 语句,这里的 break 也为 Breaks 中定义的方法。

```
def break(): Nothing = { throw breakException }
```

当 i==3 时,便结束 breakable 方法的执行,从而达到退出循环的目的。

3.3.2 有过滤条件的 for 循环

前面我们提到有两种方法可以退出 for 循环,其中一种就是通过加入 Boolean 类型的变量作为过滤条件,语法格式如下:

```
for(x <- 表达式 if 条件判断1; if 条件判断2...){
    //所有条件判断都满足时才执行循环中的语句;
}
```

其示例代码如下:

```
scala> for(i<- 1 to 5 if(i<3)) {
  println("i="+i)
}
i=1
i=2
```

for 循环表达式中加入了过滤条件 if(i<3),只有满足该条件 for 循环才会继续执行。加入多个条件判断的 for 循环示例代码如下:

```
scala> for(i<- 1 to 40 if(i%4==0);if(i%5==0)){
  println("i="+i)
}
i=20
i=40
```

3.3.3 多重 for 循环

Scala 中可以使用多重 for 循环,其语法格式如下:

```
for(x <- 表达式 if 条件判断1; if 条件判断2...){
    for(y <- 表达式 if 条件判断1; if 条件判断2...){
        //所有条件判断都满足时才执行循环中的语句;
    }
}
```

}
```

示例代码如下:

```
scala> for(i<- 1 to 5 if(i>3)){
  for(j<- 5 to 7 if(j==6)){
    println("i="+i+",j="+j)
  }
}
i=4,j=6
i=5,j=6
```

3.3.4 作为表达式的 for 循环

for 循环同关键字 yield 一起使用可作为表达式,循环执行完成后有返回值,例如:

```
scala> var x=for (i <- 1 to 5) yield i
x: scala.collection.immutable.IndexedSeq[Int] = Vector(1, 2, 3, 4, 5)
```

for 循环每执行一次,yield 就会生成对应的值并保存在缓存中。当循环执行完成后,它会利用前面缓存中得到的值生成一个集合并返回。关键字 yield 后面还可以跟表达式,例如:

```
scala> var x=for (i <- 1 to 5) yield i%2==0
x: scala.collection.immutable.IndexedSeq[Boolean] = Vector(false, true, false, true, false)

scala> var x=for (i <- 1 to 5) yield i/2
x: scala.collection.immutable.IndexedSeq[Int] = Vector(0, 1, 1, 2, 2)
```

小　结

通过本章的学习,读者应该掌握 Scala 中基本程序控制结构的使用,Scala 语言只有 if 语句、while 循环及 for 循环程序控制结构,特别需要注意 if 语句、for 循环语句具有表达式的特性,它们在运行完成后可以有返回值,而 while 循环虽然有返回值但其返回值始终为 Unit 类型。在 for 循环控制结构当中,需要注意多重 for 循环和带有过滤条件的 for 循环控制的使用。在下一章当中,将介绍 Scala 中的重要数据结构——集合。
```

第 4 章　集　合

本章将介绍 Scala 集合的整体结构、特点等，然后介绍 Scala 中常用的集合类型如数组、列表、集、映射、队列、栈并给出对应集合的常用函数的使用方法。

4.1 集合简介

Scala 中的集合（collection）分为两种，一种是可变的集合，另一种是不可变的集合。可变集合可以被更新或修改，添加、删除、修改元素将作用于原集合。而不可变集合一旦被创建，便不能被改变，添加、删除、更新操作返回的是新的集合，原有的集合保持不变。Scala 中所有的集合都来自于 scala.collection 包及其子包 mutable 和 immutable。scala.collection.immutable 包中的集合是不可变的，函数式编程语言推崇使用 immutable 集合。scala.collection.mutable 包中的集合是可变的，使用可变集合时需要开发人员明白集合何时发生变化。scala.collection 中的集合要么是 mutalbe 的，要么是 immutable 的，同时该包也定义了 immutable 及 mutable 集合的接口。

在 Scala 中，默认使用的都是 immutable 集合，如果要使用 mutable 集合，需要在程序中引入，如下述示例：

```
import scala.collection.mutable
//由于 immutable 是默认导入的，因此要使用 mutable 中的集合的话，可以使用如下语句
scala> val mutableSet=mutable.Set(1,2,3)
mutableSet: scala.collection.mutable.Set[Int] = Set(1, 2, 3)
//不指定的话，创建的是 immutable 集合
scala> val mutableSet=Set(1,2,3)
mutableSet: scala.collection.immutable.Set[Int] = Set(1, 2, 3)
```

直接使用 Set(1,2,3) 创建的是 immutable 集合，这是因为在不引入任何包的时候，scala 会默认进行下列自动引入。

```
import java.lang._
import scala._
import Predef._
```

其中 Predef 对象中包含了 Set、Map 等的定义：

```
type Map[A, +B] = immutable.Map[A, B]
type Set[A]     = immutable.Set[A]
val Map         = immutable.Map
val Set         = immutable.Set
```

这也就是代码 val mutableSet=Set(1,2,3)得到的结果是 scala.collection.immutable.Set[Int] = Set(1, 2, 3)的原因，在直接使用 Set(1,2,3)构造 Set 时，默认使用的是 Predef 对象中定义的 immutable.Set。

为便于了解 Scala 集合的整体情况，这里分别给出 scala.collection、scala.collection. immutable 及 scala.collection. immutable 包中的集合类的层次结构图。scala.collection 包中的集合类层次结构如图 4-1 所示。

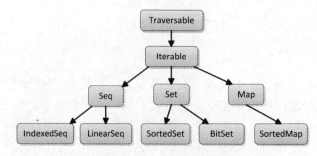

图 4-1 scala.collection 包中的集合类层次结构

scala.collection.immutable 包中的类层次结构如图 4-2 所示。

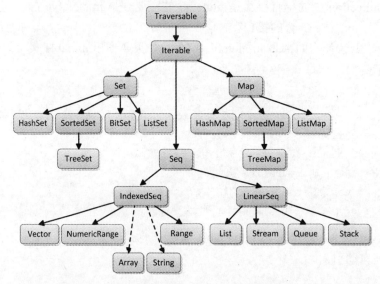

图 4-2　scala.collection.immutable 包中的类层次结构

scala.collection.mutable 包中的类层次结构如图 4-3 所示。

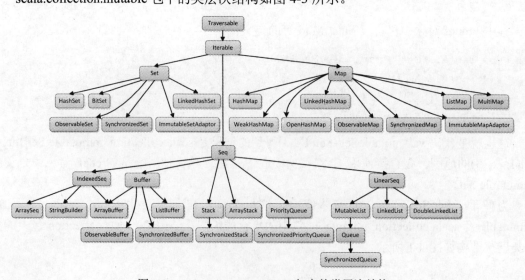

图 4-3　scala.collection.mutable 包中的类层次结构

scala.collection.mutable 和 scala.collection.immmutable 包中的常用可变集合与不可变集合存在一定的对应关系，具体如表 4-1 所示，在使用时可以根据具体场景进行选择。

表 4-1 可变集合与不可变集合对应关系

不可变（collection.immutable._）	可变（collection.mutable._）
Array	ArrayBuffer
List	ListBuffer
/	LinkedList、DoubleLinkedList
List	MutableList
/	Queue
Array	ArraySeq
Stack	Stack
HashMap	HashMap
HashSet	HashSet
	ArrayStack

4.2 数 组

数组为相同数据类型的元素按一定顺序排列的集合，在 Scala 语言中数组是最常用、最重要的数据结构。Scala 中的数组分为定长数组和变长数组，定长数组在定义时长度被确定，在运行时数组长度不会发生改变，而变长数组内存空间长度会随程序运行的需要而动态扩容。下面对定长数组和变长数组的使用、遍历方式、常用函数、多维数组等进行介绍。

4.2.1 定长数组

定长数组指的是数组长度在定义时被确定，数组占有的内存空间在程序运行时不会被改变，定长数组的定义如下：

```
//定义一个长度为 10 的数值数组
scala> val numberArray=new Array[Int](10)
numberArray: Array[Int] = Array(0, 0, 0, 0, 0, 0, 0, 0, 0, 0)
//定义一个长度为 10 的 String 类型数组
scala> val strArray=new Array[String](10)
strArray: Array[String] = Array(null, null, null, null, null, null, null, null, null, null)
```

通过上述代码可以看到，非数值对象类型在数组定义时被初始化为 null，数值对象类型被初始化为 0。定义完数组后可以对数组内容进行访问，具体示例代码如下：

```
//数组元素赋值
scala> strArray(0)="First Element"
```

```
//需要注意的是，val strArray=new Array[String](10)
//这意味着 strArray 不能被改变，但数组内容是可以改变的
scala> strArray
res62: Array[String] = Array(First Element, null, null, null, null, null, null, null, null, null)
```

前面 String 类型的数组 strArray、Int 类型的数组 numberArray，都是直接通过显式地 new 创建，在 Scala 中还有一种无 new 操作的数组定义方式，这种方式可以在定义数组时直接对数组元素内容进行赋值，例如：

```
//另一种定长数组定义方式，这种定义方式其实是调用其 apply 方法进行数组创建操作
scala> val strArray2=Array("First","Second")
strArray2: Array[String] = Array(First, Second)
```

直接使用代码 Array("First","Second")创建数组并对数组元素进行赋值，事实上它调用的是 Array 伴生对象的 apply 方法，这种集合创建方式在后面还经常会遇到。对于伴生对象的 apply 方法，我们会在 Scala 面向对象编程部分作进一步介绍。

4.2.2 变长数组 ArrayBuffer

变长数组在程序运行过程中，其数组长度可以随程序运行的需要而增加。最常用的变长数组为 ArrayBuffer，它在包 scala.collection.mutable 中被使用时需要显式地引入，具体示例代码如下：

```
//要使用 ArrayBuffer，先要引入 scala.collection.mutable.ArrayBuffer
scala> import scala.collection.mutable.ArrayBuffer
import scala.collection.mutable.ArrayBuffer

//创建 String 类型 ArrayBuffer 数组缓冲
scala> val strArrayVar=ArrayBuffer[String]()
strArrayVar: scala.collection.mutable.ArrayBuffer[String] = ArrayBuffer()
```

上述代码创建了 String 类型的变长数组，同样也可以直接通过类名 ArrayBuffer[String]()这样的方式创建。ArrayBuffer 来自 scala.collection.mutable 包，当在数组中进行添加、删除、修改等操作时，它直接作用于原始定义的集合，示例代码如下：

```
//+=意思是在尾部添加元素
scala>    strArrayVar+="Hello"
res63: strArrayVar.type = ArrayBuffer(Hello)

//+=后面还可以跟多个元素的集合
//注意操作后的返回值
scala> strArrayVar+=("World","Programmer")
res64: strArrayVar.type = ArrayBuffer(Hello, World, Programmer)
```

```
//显示完整的数组内容
scala> strArrayVar
res65: scala.collection.mutable.ArrayBuffer[String] = ArrayBuffer(Hello, World,Programmer)

//++=用于向数组中追加内容，++=右侧可以是任何集合
//追加 Array 数组
scala> strArrayVar++=Array("Wllcome","To","Scala World")
res66: strArrayVar.type = ArrayBuffer(Hello, World, Programmer, Wllcome, To, Scala World)

//追加 List
scala> strArrayVar++=List("Wellcome","To"," Scala World ")
res67: strArrayVar.type = ArrayBuffer(Hello, World, Programmer, Wllcome, To, Scala World, Wellcome, To, Scala World)

//删除末尾 n 个元素
scala> strArrayVar.trimEnd(3)

scala> strArrayVar
res69: scala.collection.mutable.ArrayBuffer[String] = ArrayBuffer(Hello, World,Programmer, Wllcome, To, Scala World)
```

前面给出的是 String 类型的 ArrayBuffer 的定义及操作。接下来我们给出 Int 类型的 ArrayBuffer 定义及相关操作，示例如下：

```
//创建整型数组缓冲
scala> var intArrayVar=ArrayBuffer(1,1,2)
intArrayVar: scala.collection.mutable.ArrayBuffer[Int] = ArrayBuffer(1, 1, 2)

//在数组索引为 0 的位置插入元素 6
scala> intArrayVar.insert(0,6)

scala> intArrayVar
res72: scala.collection.mutable.ArrayBuffer[Int] = ArrayBuffer(6, 1, 1, 2)

//在数组索引为 0 的位置插入元素 7,8,9
scala> intArrayVar.insert(0,7,8,9)

scala> intArrayVar
res74: scala.collection.mutable.ArrayBuffer[Int] = ArrayBuffer(7, 8, 9, 6, 1, 1, 2)

//从索引 0 开始，删除 4 个元素
```

```
scala> intArrayVar.remove(0,4)

scala> intArrayVar
res77: scala.collection.mutable.ArrayBuffer[Int] = ArrayBuffer(1, 1, 2)
```

变长数组 ArrayBuffer 与定长数组 Array 还可以相互转换，例如：

```
//转成定长数组
scala> intArrayVar.toArray
res78: Array[Int] = Array(1, 1, 2)

//将定长数组转成 ArrayBuffer
scala> res78.toBuffer
res80: scala.collection.mutable.Buffer[Int] = ArrayBuffer(1, 1, 2)
```

4.2.3 数组的遍历

在上一章中我们已经介绍了 for 循环的使用方法，而集合遍历使用的都是 for 循环，数组遍历也不例外。数组的遍历有两种方式：通过索引遍历和直接数组遍历。

下面的代码演示的是索引遍历数组的方式。

```
//to
scala> for(i <- 0 to intArrayVar.length-1) println("Array Element: " +intArrayVar(i))
Array Element: 1
Array Element: 1
Array Element: 2

//until
scala> for(i <- 0 until intArrayVar.length) println("Array Element: " +intArrayVar(i))
Array Element: 1
Array Element: 1
Array Element: 2
```

索引遍历数组的好处是可以在 for 循环中加入过滤条件从而访问特定的数组元素，具体示例代码如下：

```
//步长为 2
scala>  for(i <- 0 until (intArrayVar.length,2)) println("Array Element: " +intArrayVar(i))
Array Element: 1
Array Element: 2
```

```
//倒序输出
scala> for( i<- (0 until intArrayVar.length).reverse) println("Array Element: "+
 intArrayVar(i))
Array Element: 2
Array Element: 1
Array Element: 1
```

不过在实际开发中,还是推荐直接数组遍历的方式,具体示例代码如下:

```
//数组方式(推荐使用)
scala> for(i <- intArrayVar) println("Array Element: " + i)
Array Element: 1
Array Element: 1
Array Element: 2
```

4.2.4 遍历生成数组

无论是定长数组还是变长数组,在遍历的时候都可以生成新的数组,生成新数组时,原来的数组内容保持不变,具体示例代码如下:

```
//生成新的数组,原数组不变,变长数组遍历生成的数组仍然是变长数组
scala> var intArrayVar2=for(i <- intArrayVar) yield i*2
intArrayVar2: scala.collection.mutable.ArrayBuffer[Int] = ArrayBuffer(2, 2, 4)

//定长数组遍历生成后的数组仍然是定长数组,原数组不变
scala> var intArrayNoBuffer=Array(1,2,3)
intArrayNoBuffer: Array[Int] = Array(1, 2, 3)

scala> var intArrayNoBuffer2=for(i <- intArrayNoBuffer) yield i*2
intArrayNoBuffer2: Array[Int] = Array(2, 4, 6)

//加入过滤条件
scala>  var intArrayNoBuffer2=for(i <- intArrayNoBuffer if i>=2) yield i*2
intArrayNoBuffer2: Array[Int] = Array(4, 6)
```

4.2.5 常用函数

数组中定义了有大量的函数,本节只给出一般函数的使用方法,至于数组中的 map、filter 等高阶函数的使用,我们将在第 5 章中介绍。常用数组操作函数使用示例如下:

```
//定义一个整型数组
scala> val intArr=Array(1,2,3,4,5,6,7,8,9,10)
intArr: Array[Int] = Array(1, 2, 3, 4, 5, 6, 7, 8, 9, 10)
```

```
//求和
scala> intArr.sum
res87: Int = 55

//求最大值
scala> intArr.max
res88: Int = 10

scala> ArrayBuffer("Hello","Hell","Hey","Happy").max
res90: String = Hey

//求最小值
scala> intArr.min
res89: Int = 1

//toString()方法
scala> intArr.toString()
res94: String = [I@141aba8

//mkString()方法
scala> intArr.mkString(",")
res96: String = 1,2,3,4,5,6,7,8,9,10

scala> intArr.mkString("<",",",">")
res97: String = <1,2,3,4,5,6,7,8,9,10>
```

4.3.6 多维数组

实现多维数组的定义如下:

```
//定义2行3列数组
scala> var multiDimArr=Array(Array(1,2,3),Array(2,3,4))
multiDimArr: Array[Array[Int]] = Array(Array(1, 2, 3), Array(2, 3, 4))
```

上述代码定义了一个2行3列的数组。可以通过索引访问数组元素,例如:

```
//获取第一行第三列元素
scala> multiDimArr(0)(2)
res99: Int = 3
```

也可通过for循环遍历数组,示例如下:

```
//多维数组的遍历
scala> for(i <- multiDimArr) println( i.mkString(","))
1,2,3
2,3,4
```

```
//通过双重for循环遍历多维数组
scala> for(i<- multiDimArr)
  for(j<- i) print(j+" ")
1 2 3 2 3 4
```

对于任意维的数组,都可以通过多重for循环来达到遍历数组的目的。

4.3 列　表（List）

同数组一样,List也是Scala语言中应用十分广泛的集合类型数据结构,主要存在两种类型的List,分别是scala.collection.immutable.List[A]及scala.collection.mutable包中的DoubleLinkedList、LinkedList及ListBuffer等。本节只介绍scala.collection.immutable.List[A]及其使用方法,它的类继承层次结构如图4-4所示。

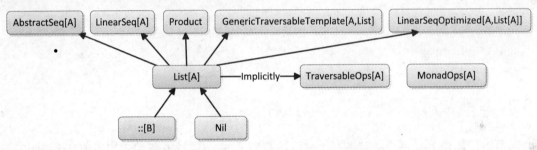

图4-4　列表(List)类继承层次结构

通过上图可以看到,List有一个子类分别是::[B]及一个继承List的对象Nil,其中::表示的是头尾相接的非空列表,而Nil表示的是空列表。

4.3.1　列表的创建

列表的创建如下所示:

```
//字符串类型List
scala> val listStr=List("Spark","Hive","Flink")
listStr: List[String] = List(Spark, Hive, Flink)

//前一个语句与下面语句等同
scala> val listStr=List.apply("Spark","Hive","Flink")
listStr: List[String] = List(Spark, Hive, Flink)

//数值类型List
scala> val doubleList=List(1.0,2.0,3.0)
doubleList: List[Double] = List(1.0, 2.0, 3.0)
```

```
//多重List, List的子元素为List
scala> val multiDList=List(List(1,2,3),List(4,5,6),List(7,8,9))
multiDList: List[List[Int]] = List(List(1, 2, 3), List(4, 5, 6), List(7, 8, 9))

//遍历List
scala> for(i<- multiDList) println(i)
List(1, 2, 3)
List(4, 5, 6)
List(7, 8, 9)
```

List 与 Array 有很多相似之处。除了通过 List("Spark","Hive","Flink") 及 List.apply("Spark","Hive","Flink")等方式创建List之外，还可以使用::操作符创建List。例如：

```
//采用::及Nil进行列表构建
scala> val nums = 1 :: (2 :: (3 :: (4 :: Nil)))
nums: List[Int] = List(1, 2, 3, 4)

//由于::操作符的优先级是从右往左的，因此上一条语句等同于下面这条语句
scala> val nums=1::2::3::4::Nil
nums: List[Int] = List(1, 2, 3, 4)
```

4.3.2 List常用函数

本节给出的是 List 常用函数，对于高阶函数如 map、flatMap 等的使用将放在 Scala 函数部分进行介绍。

```
//判断是否为空
scala> nums.isEmpty
res108: Boolean = false

//取第一个元素
scala> nums.head
res109: Int = 1

//取除第一个元素外剩余的元素，返回的是列表
scala> nums.tail
res114: List[Int] = List(2, 3, 4)

//取列表第二个元素
scala> nums.tail.head
res115: Int = 2

//List连接操作
scala> List(1,2,3):::List(4,5,6)
```

```
res116: List[Int] = List(1, 2, 3, 4, 5, 6)

//去除最后一个元素外的元素，返回的是列表
scala> nums.init
res117: List[Int] = List(1, 2, 3)

//取列表最后一个元素
scala> nums.last
res118: Int = 4

//列表元素倒置
scala> nums.reverse
res119: List[Int] = List(4, 3, 2, 1)

//一些好玩的方法调用
scala> nums.reverse.reverse==nums
res120: Boolean = true

scala> nums.reverse.init
res121: List[Int] = List(4, 3, 2)

scala> nums.tail.reverse
res122: List[Int] = List(4, 3, 2)

//丢弃前n个元素
scala> nums drop 3
res123: List[Int] = List(4)

scala> nums drop 1
res124: List[Int] = List(2, 3, 4)

//获取前n个元素
scala> nums take 1
res125: List[Int] = List(1)

scala> nums.take(3)
res126: List[Int] = List(1, 2, 3)

//将列表进行分割
scala> nums.splitAt(2)
res127: (List[Int], List[Int]) = (List(1, 2),List(3, 4))

//上面操作与下列语句等同
scala> (nums.take(2),nums.drop(2))
```

```
res128: (List[Int], List[Int]) = (List(1, 2),List(3, 4))

//Zip 操作
scala> val nums=List(1,2,3,4)
nums: List[Int] = List(1, 2, 3, 4)

scala> val chars=List('1','2','3','4')
chars: List[Char] = List(1, 2, 3, 4)

//返回的是 List 类型的元组(Tuple)
scala> nums zip chars
res130: List[(Int, Char)] = List((1,1), (2,2), (3,3), (4,4))

//List toString 方法
scala> nums.toString
res131: String = List(1, 2, 3, 4)

//List mkString 方法
scala> nums.mkString
res132: String = 1234

//转换成数组
scala> nums.toArray
res134: Array[Int] = Array(1, 2, 3, 4)
```

4.3.3 List 伴生对象方法

List 伴生对象中还存在一些常用的 List 创建方法。伴生对象中的方法可以直接通过伴生对象名称.伴生对象方法的方式进行调用，具体示例代码如下：

```
//apply 方法
scala> List.apply(1, 2, 3)
res139: List[Int] = List(1, 2, 3)

//range 方法，构建某一值范围内的 List
scala> List.range(2, 6)
res140: List[Int] = List(2, 3, 4, 5)

//步长为 2
scala> List.range(2, 6,2)
res141: List[Int] = List(2, 4)

//步长为-1
scala> List.range(2, 6,-1)
```

```
res142: List[Int] = List()

scala> List.range(6,2 ,-1)
res143: List[Int] = List(6, 5, 4, 3)

//构建相同元素的List
scala> List.make(5, "hey")
res144: List[String] = List(hey, hey, hey, hey, hey)

//unzip方法
scala> List.unzip(res145)
res146: (List[Int], List[Char]) = (List(1, 2, 3, 4),List(1, 2, 3, 4))

//list.flatten,将列表平滑成第一个无素
scala> val xss =List(List('a', 'b'), List('c'), List('d', 'e'))
xss: List[List[Char]] = List(List(a, b), List(c), List(d, e))
scala> xss.flatten
res147: List[Char] = List(a, b, c, d, e)

//列表连接
scala> List.concat(List('a', 'b'), List('c'))
res148: List[Char] = List(a, b, c)
```

4.4 集（Set）

Set 是一种不存在重复元素的集合，它与数学上定义的集合是对应的。图 4-5 给出了 Set 类的继承层次结构。

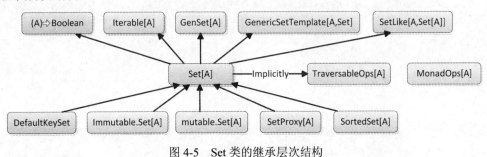

图 4-5 Set 类的继承层次结构

通过图 4-5 可以看到，Set 主要有两种类型：immutable.Set[A]和 mutable.Set[A]，其中 immutable.Set 继承层次结构如图 4-6 所示。

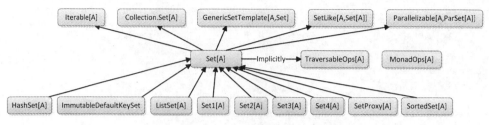

图 4-6 immutable.Set 继承层次结构

而 mutable.Set 继承层次结构如图 4-7 所示。

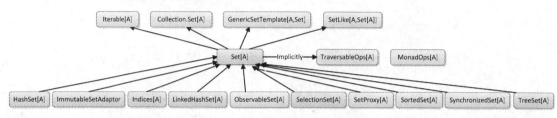

图 4-7 mutable.Set 继承层次结构

下面简单演示 scala.collection.mutable.Set 及 scala.collection.mutalbe.LinkedHashSet 的使用,其他类型的 Set 使用可以参照官方文档。

```
scala> import scala.collection.mutable.Set
import scala.collection.mutable.Set

//定义一个集合,这里使用的是mutable
scala> val numsSet=Set(3.0,5)
numsSet: scala.collection.mutable.Set[Double] = Set(5.0, 3.0)

//向集中添加一个元素,同前面的列表和数组不一样的是,Set 在插入元素时并不保证元素的顺序
//默认情况下,Set 的实现方式是 HashSet 实现方式,集中的元素通过 HashCode 值进行组织
scala> numsSet+6
res20: scala.collection.mutable.Set[Double] = Set(5.0, 6.0, 3.0)

//遍历集
scala> for ( i <- res20 ) println(i)
5.0
6.0
3.0

//如果对插入的顺序有着严格的要求,则采用 scala.collection.mutalbe.LinkedHashSet 来实现
scala> val linkedHashSet=scala.collection.mutable.LinkedHashSet(3.0,5)
linkedHashSet: scala.collection.mutable.LinkedHashSet[Double] = Set(3.0, 5.0)

scala> linkedHashSet+6
```

```
res26: scala.collection.mutable.LinkedHashSet[Double] = Set(3.0, 5.0, 6.0)
```

4.5 映射（Map）

Map 是一种键值对的集合，常被译为映射。Scala 中的 Map 类型有很多种，其继承层次结构如图 4-8 所示。

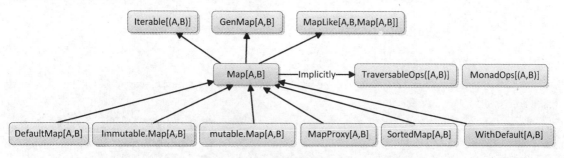

图 4-8　Map 类型继承层次结构

可以看到，Map 也有两种类型，分别是 immutable.Map[A,B]、mutable.Map[A,B]，其中 immutable.Map[A,B]的类继承层次结构如图 4-9 所示。

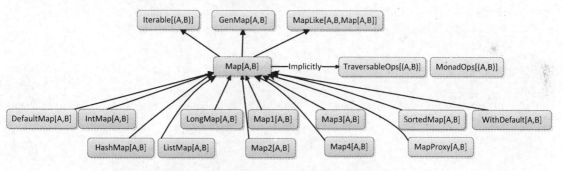

图 4-9　immutable.Map[A,B]的类继承层次结构

mutable.Map[A,B]对应的类继承层次结构如图 4-10 所示。

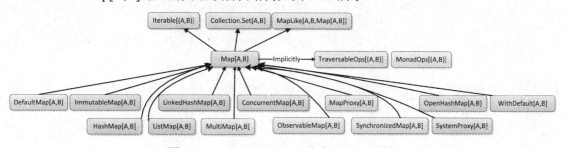

图 4-10　mutable.Map[A,B]的类继承层次结构

下面简单演示 scala.collection.immutable.Map 及 scala.collection.mutalbe.Map 的使用，其他类型的 Map 使用可以参照官方文档

```
//直接初始化，->操作符，左边是key,右边是value
```

```
scala> val studentInfo=Map("john" -> 21, "stephen" -> 22,"lucy" -> 20)
studentInfo: scala.collection.immutable.Map[String,Int] = Map(john -> 21, stephen -> 22, lucy -> 20)

//immutable 不可变，它不具有以下操作
scala> studentInfo.clear()
<console>:10: error: value clear is not a member of scala.collection.immutable.M
ap[String,Int]
              studentInfo.clear()
                          ^
//创建可变的 Map
scala> val studentInfoMutable=scala.collection.mutable.Map("john" -> 21, "stephen" -> 22,"lucy" -> 20)
studentInfoMutable: scala.collection.mutable.Map[String,Int] = Map(john -> 21, lucy -> 20, stephen -> 22)

//mutable Map 可变，比如可以将其内容清空
scala> studentInfoMutable.clear()

scala> studentInfoMutable
res3: scala.collection.mutable.Map[String,Int] = Map()

scala> val studentInfoMutable=scala.collection.mutable.Map("john" -> 21, "stephen" -> 22,"lucy" -> 20)
studentInfoMutable: scala.collection.mutable.Map[String,Int] = Map(john -> 21, lucy -> 20, stephen -> 22)

//遍历操作 1
scala> for( i <- studentInfoMutable ) println(i)
(john,21)
(lucy,20)
(stephen,22)

//遍历操作 2
scala> studentInfoMutable.foreach(e=>{val (k,v)=e; println(k+":"+v)}
     )
john:21
lucy:20
stephen:22

//遍历操作 3
scala> studentInfoMutable.foreach(e=> println(e._1+":"+e._2))
john:21
```

```
lucy:20
stephen:22

//定义一个空的HashMap
scala> val xMap=new scala.collection.mutable.HashMap[String,Int]()
xMap: scala.collection.mutable.HashMap[String,Int] = Map()

//往里面填充值
scala> xMap.put("spark",1)
res12: Option[Int] = None

scala> xMap
res13: scala.collection.mutable.HashMap[String,Int] = Map(spark -> 1)

//判断是否包含spark字符串
scala> xMap.contains("spark")
res14: Boolean = true

//-> 初始化Map,也可以通过("spark",1)这种方式实现(元组的形式)
scala> val xMap=scala.collection.mutable.Map(("spark",1),("hive",1))
xMap: scala.collection.mutable.Map[String,Int] = Map(spark -> 1, hive -> 1)

scala> "spark" -> 1
res18: (String, Int) = (spark,1)

//获取元素
scala> xMap.get("spark")
res19: Option[Int] = Some(1)

scala> xMap.get("SparkSQL")
res20: Option[Int] = None
```

4.6 队 列（Queue）

队列在现代编程语言中同样是一种非常重要的数据结构，Scala 中的队列有两种，分别是 scala.collection.mutable.Queue 和 scala.collection.immutable.Queue，其中 mutable.Queue 的类继承层次结构如图 4-11 所示。

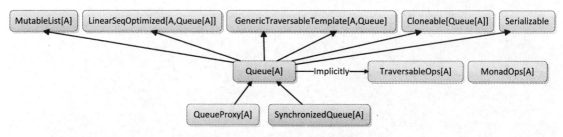

图 4-11 mutable.Queue 的类继承层次结构

下面给出 scala.collection.immutable.Queue 的简单使用，更多方法及 scala.collection.mutable.Queue 的使用可以查看 scala 的官方文档。

```
//immutable queue
scala> var queue=scala.collection.immutable.Queue(1,2,3)
queue: scala.collection.immutable.Queue[Int] = Queue(1, 2, 3)

//出队
scala> queue.dequeue
res38: (Int, scala.collection.immutable.Queue[Int]) = (1,Queue(2, 3))

//入队
scala> queue.enqueue(4)
res40: scala.collection.immutable.Queue[Int] = Queue(1, 2, 3, 4)

//mutable queue
scala> var queue=scala.collection.mutable.Queue(1,2,3,4,5)
queue: scala.collection.mutable.Queue[Int] = Queue(1, 2, 3, 4, 5)

//入队操作
scala> queue += 5
res43: scala.collection.mutable.Queue[Int] = Queue(1, 2, 3, 4, 5, 5)

//集合方式
scala> queue ++= List(6,7,8)
res45: scala.collection.mutable.Queue[Int] = Queue(1, 2, 3, 4, 5, 5, 6, 7, 8)
```

4.7 栈（Stack）

Scala 中也提供了栈的实现，Scala 中的栈有两种，分别是 scala.collection.mutable.Stack 和 scala.collection.immutable.Stack，其中 mutable.Stack 的类继承层次结构图 4-12 所示。

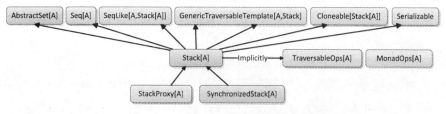

图 4-12 mutable.Stack 的类继承层次结构

下面给出的是 scala.collection.mutable.Stack 的常见函数使用，至于其子类 SynchronizedStack 及不可变的栈 scala.collection.immutable.Stack 的使用可以查看 Scala 官方文档。

```
//mutable Stack
scala> import scala.collection.mutable.Stack
import scala.collection.mutable.Stack

//new 创建方式
scala> val stack = new Stack[Int]
stack: scala.collection.mutable.Stack[Int] = Stack()

//Apply 创建方式
scala> val stack1=Stack(1,2,3)
stack1: scala.collection.mutable.Stack[Int] = Stack(1, 2, 3)

//出栈
scala> stack1.top
res55: Int = 1

//入栈
scala> stack.push(1)
res57: stack.type = Stack(1)

//入栈
scala> stack.push(2)
res58: stack.type = Stack(2, 1)

//出栈
scala> stack.top
res59: Int = 2

scala> stack
res60: scala.collection.mutable.Stack[Int] = Stack(2, 1)
```

小 结

　　本章首先对 Scala 集合的整体情况进行了介绍,然后给出数组(Array)、列表(List)、集(Set)、映射(Map)、队列(Queue)及栈(Stack)的创建、基本函数的使用。在下一章中,我们将介绍 Scala 的函数式编程。

第 5 章　函　数

本章将介绍 Scala 函数的定义与使用、函数字面量的定义与使用、高阶函数的使用、闭包的定义与使用、函数的柯里化、部分应用函数及偏函数的使用等内容。

5.1 函 数

在 Scala 中，一个标准的函数定义如图 5-1 所示。

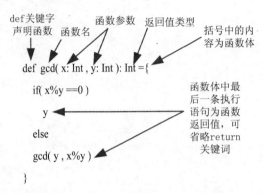

图 5-1 函数定义

图中 def 关键字用于声明一个函数，gcd 为函数名称，函数名称与变量名的定义类似，可以是任意的合法字符串，(x:Int,y:Int)为函数参数，函数参数后面的:Int 用于指定函数的返回值类型，=号后面是函数体，函数体如果是多行语句，则需要将其放在大括号中，如果只有一条语句，则可以省略。

```
//函数体中只有一行语句时，可以省略{}
scala> def gcd(x:Int,y:Int):Int=if(x%y==0) y else gcd(y,x%y)
gcd: (x: Int, y: Int)Int
```

函数中最后一条执行语句为函数的返回值，函数返回值可以加 return 关键词，也可以将其省略。

```
//return 关键字也可以不省略
scala> def gcd(x:Int,y:Int):Int={
  if(x%y==0)
    return y
  else
    return gcd(y,x%y)
}
gcd: (x: Int, y: Int)Int
```

Scala 具有类型推导功能，会根据最终的返回值推导函数的返回值类型，因此在实际应用中也常常会省略函数的返回值。

```
//省略函数返回值，Scala 会通过类型推导来确定函数的返回值类型
scala> def sum(x:Int,y:Int)=x+y
sum: (x: Int, y: Int)Int
```

不过，类型推导有两个限制：

（1）如果需要 return 关键字指定返回值，则必须显式地指定函数返回值的类型。

```
scala> def sum(x:Int,y:Int)=return x+y
<console>:7: error: method sum has return statement; needs result type
       def sum(x:Int,y:Int)=return x+y
```

（2）如果函数中存在递归调用，则必须显式地指定函数返回值的类型。

```
scala> def gcd(x:Int,y:Int)={
  if(x%y==0)
    y
  else
    gcd(y,x%y)
}
<console>:12: error: recursive method gcd needs result type
           gcd(y,x%y)
```

5.2 值函数

5.2.1 值函数的定义

在 Scala 语言中函数也是对象，也可以像变量一样被赋值，把这种函数称为函数字面量（function literal）或值函数，标准的函数字面量定义如图 5-2 所示。

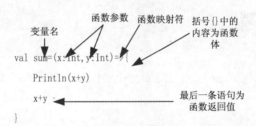

图 5-2 标准函数字面量定义

图中的(x:Int,y:Int)=>{ println(x+y) x+y }定义了一个没有名字的函数，然后将其赋值给变量 val sum。可以看到赋值时，如同一个普通的变量赋值一样。代码(x:Int,y:Int)=>{ println(x+y) x+y }中的(x:Int,y:Int)为值函数的输入参数，=>为函数映射符，意思是将=>左边的某种类型转换成=>右边的某种类型（本例中为将 Int 类型转换成 Int 类型），{ println(x+y) x+y }为函数体，多行语句必须放置在{}中，如果只有一行语句，则可以省略，具体示例如下：

```
//一行语句时的值函数
scala>val sum=(x:Int,y:Int)=>x+y
sum: (Int, Int) => Int = <function2>
```

形如(x:Int,y:Int)=>{ println(x+y); x+y }的表达式被称为 Lambda 表达式，因此值函数 val sum=(x:Int,y:Int)=>{ println(x+y); x+y }常被称为 Lambda 函数。

使用值函数需要特别注意的是，值函数不能像普通函数那样指定其返回值类型，编译器

会进行类型推导来确定函数返回值类型,示例代码如下:

```
//值函数不能在参数后面指定返回值类型
scala> val sum=(x:Int,y:Int):Int=>x+y
<console>:7: error: not found: type +
       val sum=(x:Int,y:Int):Int=>x+y

<console>:7: error: not found: value x
       val sum=(x:Int,y:Int):Int=>x+y

<console>:7: error: not found: value y
       val sum=(x:Int,y:Int):Int=>x+y
```

值函数在定义时会被编译成一个类,为对此说明,我们定义下列应用程序对象:

```
/*
 * 值函数
 */
object Example5_2 extends App{
  //多行语句时
  val sum=(x:Int,y:Int)=>{
    println(x+y)
    x+y
  }
  //一行语句时的函数字面量
  //val sum=(x:Int,y:Int):Int=>x+y
}
```

编译完成后会生成名为 Example5_2$$anonfun$1.class 的字码文件,对其反编译,可以得到如下代码:

```
E:\IntellijIDEA14Workspace\ScalaProject\out\production\ScalaProject\cn\sca
la\chapter04>javap -private Example5_2$$anonfun$1.class
Compiled from "chapter04.scala"
public final class cn.scala.chapter04.Example5_2$$anonfun$1 extends scala.r
untim
e.AbstractFunction2$mcIII$sp implements scala.Serializable {
  public static final long serialVersionUID;
  public final int apply(int, int);
  public int apply$mcIII$sp(int, int);
  public final java.lang.Object apply(java.lang.Object, java.lang.Object);
  public cn.scala.chapter04.Example5_2$$anonfun$1();
}
```

可以看到,类 Example5_2$$anonfun$1 继承了一个名为 scala.runtime.AbstractFunction2$mcIII$sp 的函数类,在使用时,该值函数被创建或成为一个对象,然后赋值给相应的变量。

值函数还可以如一般的 val 变量一样使用 lazy 关键字修饰,使用 lazy 修饰的函数字面量的作用原理同一般的 val 变量一样,只有当程序真正使用到该函数字面量的时候才会被创建,例如:

```
scala> val f=(x:Double)=>x*2
f: Double => Double = <function1>

scala> f(10)
res4: Double = 20.0

//使用 lazy 关键字修饰函数字面量,当真正使用到该变量的时候,函数才被创建
scala> lazy val f=(x:Double)=>x*2
f: Double => Double = <lazy>

scala> f(10)
res7: Double = 20.0

scala> f
res8: Double => Double = <function1>
```

5.2.2 值函数的简化

值函数最常用的场景是作为高阶函数的输入,假设定义了如下整型数组:

```
scala> val arrInt=Array(1,2,3,4)
arrInt: Array[Int] = Array(1, 2, 3, 4)
```

现在希望能够将数组中所有的元素加 1,可以通过 Array 提供的 map 方法来处理,map 方法的定义如下所示:

```
def map[B](f: (A)    B): Array[B]
```

该方法使用了泛型,泛型 B 指的是数组最终返回的元素类型,f 表示作用于数组各元素上的函数,(A) ⇒ B 为函数的类型,表示的是该函数的输入参数类型为 A,返回值类型为 B。这里我们定义如下值函数作为 map 函数的输入。

```
scala> val increment=(x:Int)=>x+1
increment: Int => Int = <function1>
```

使用时直接将变量 increment 作为函数的参数使用。

```
scala> arrInt.map(increment)
res0: Array[Int] = Array(2, 3, 4, 5)
```

在实际使用时,如果值函数只使用一次,则常常直接作为函数参数。

```
scala> arrInt.map((x:Int)=>x+1)
```

```
res1: Array[Int] = Array(2, 3, 4, 5)
```

在实际使用时,还可以对值函数进行简化,对于代码 arrInt.map((x:Int)=>x+1),由于 arrInt 为整型数组,因此传入 map 的函数输入类型可以通过类型推断得到,arrInt.map((x:Int)=>x+1) 可以简写为如下所示:

```
//值函数省略参数类型,参数类型通过类型推断得到
scala> arrInt.map((x)=>x+1)
res2: Array[Int] = Array(2, 3, 4, 5)
```

arrInt.map((x)=>x+1)中的值函数(x)=>x+1 参数只有一个,在这种条件下,可以将值函数输入参数的括号去掉。

```
//值函数参数只有一个时,可以省略括号
scala> arrInt.map(x=>x+1)
res3: Array[Int] = Array(2, 3, 4, 5)
```

代码 arrInt.map(x=>x+1)中的值函数 x=>x+1 输入参数在符号=>右边只出现一次,此时可以用占位符"_"对该值函数进行进一步简化。

```
//值函数参数在=>右边只出现一次时,可以用占位符"_"对函数进行简化
scala> arrInt.map(_+1)
res4: Array[Int] = Array(2, 3, 4, 5)
```

需要特别注意的是,代码 arrInt.map(_+1)合法是因为 arrInt 是 Int 类型的数组,可以进行类型推断,但如果将_+1 赋值给变量时会出错。

```
scala> val increment=1+_
<console>:7: error: missing parameter type for expanded function ((x$1) => 1.+(x$1))
       val increment=1+_
                       ^

scala> val increment= _+1
<console>:7: error: missing parameter type for expanded function ((x$1) => x$1.$plus(1))
       val increment= _+1
                      ^
```

不管是 val increment=1+_还是 val increment=_+1,其出错原因是编译器无法进行类型推断,虽然代码中有+1 或 1+这种表达式,但它不足以让编译器知道确定的类型,因为无论是 1+字符串类型还是字符串类型+1,1+Double 类型还是 Double 类型+1 等都是合法的。要合法使用上面的代码,需要明确指定类型。

```
//明确指定类型
scala> val increment=(_:Int)+1
increment: Int => Int = <function1>
```

```
scala> val increment=1+(_:Int)
increment: Int => Int = <function1>
```

还有一种方式，是指定 increment 的具体类型。

```
//指定 increment 的参数类型为(Int)=>Int,即函数的输入参数为 Int 类型,返回值的类型为 Int
scala> val increment:(Int)=>Int=1+_
increment: Int => Int = <function1>
```

5.3 高阶函数

5.3.1 高阶函数的定义

在前面的例子中，我们已经接触过了高阶函数，如 arrInt.map(x=>x+1)中调用数组的 map 函数，map 函数的输入参数类型为函数类型，因此 map 便是高阶函数。

首先来看看高阶函数是如何定义的。

```
//定义一个高阶函数，该函数的输入参数为函数类型(Double)=>Double
scala> def higherOrderFunction(f:(Double)=>Double)=f(100)
higherOrderFunction: (f: Double => Double)Double

//定义一个函数，该函数输入参数类型为 Double,返回值类型为 Double
scala> def sqrt(x:Double)=Math.sqrt(x)
sqrt: (x: Double)Double

//将 sqrt 函数作为高阶函数 higherOrderFunction 的输入
scala> higherOrderFunction(sqrt)
res0: Double = 10.0
```

代码 def higherOrderFunction(f:(Double)=>Double)=f(100) 定义了一个高阶函数 higherOrderFunction，该函数的参数类型为函数类型(Double)=>Double，即传入的函数要求其输入类型是 Double 类型，返回值类型也为 Double。def sqrt(x:Double)=Math.sqrt(x)定义了这样一个满足输入类型是 Double，返回值类型也是 Double 的函数，然后将 sqrt 函数作为 higherOrderFunction 的参数传入：higherOrderFunction(sqrt)。

高阶函数除能够将函数作为函数参数外，还可以将函数作为返回值，例如：

```
scala> def higherOrderFunction(factor:Int):Double=>Double={
  println("返回新的函数")
  (x:Double)=>factor*x
}
higherOrderFunction: (factor: Int)Double => Double
```

代码 def higherOrderFunction(factor:Int):Double=>Double 定义了一个函数，该函数输入参

数为 Int 类型，返回值为函数类型 Double=>Double，返回的函数类型输入参数为 Double，返回值类型为 Double，代码(x:Double)=>factor*x 为该函数的返回值。同一般的类型一样，Scala 也可以进行函数类型推导。

```
//函数类型推导，返回的函数类型为 Double => Double
scala> def higherOrderFunction(factor:Int)=(x:Double)=>factor*x
higherOrderFunction: (factor: Int)Double => Double

//生成新的函数，函数参数类型为 Double，返回值类型为 Double
scala> val multiply=higherOrderFunction(100)
multiply: Double => Double = <function1>

//调用函数
scala> multiply(10)
res0: Double = 1000.0
```

5.3.2 高阶函数的使用

通过上一小节，我们已经清楚了高阶函数的定义及使用方法，本节将以数组（Array）中的高阶函数为例介绍集合中常见高阶函数的使用。

（1）map 函数

定义：def map[B](f: (A) ⇒ B): Array[B]

用途：将函数 f 应用于数组的所有元素，并返回一个新的数组 Array[B]。函数 f 输入类型为 A、返回值类型为 B。

在介绍值函数时，已经对 Array 类型的 map 函数使用作了介绍，如下所示。

```
//Array 类型的 map 函数使用
scala> Array("spark","hive","hadoop").map(_*2)
res1: Array[String] = Array(sparkspark, hivehive, hadoophadoop)
```

下面再分别给出 List 类型、Map 类型的 map 函数使用示例，希望能够帮助读者对高阶函数 map 有更深入的理解。

```
//List 类型的 map 函数使用
scala> val list=List("Spark"->1,"hive"->2,"hadoop"->2)
list: List[(String, Int)] = List((Spark,1), (hive,2), (hadoop,2))

//省略值函数的输入参数类型
scala> list.map(x=>x._1)
res20: List[String] = List(Spark, hive, hadoop)

//参数 x 在=>中只出现一次，进一步简化
scala> list.map(_._1)
```

```
res21: List[String] = List(Spark, hive, hadoop)

//Map 类型的 map 函数使用
scala> Map("spark"->1,"hive"->2,"hadoop"->3).map(_._1)
res23: scala.collection.immutable.Iterable[String] = List(spark, hive, hadoop)

scala> Map("spark"->1,"hive"->2,"hadoop"->3).map(_._2)
res24: scala.collection.immutable.Iterable[Int] = List(1, 2, 3)

//非简化写法
scala> Map("spark"->1,"hive"->2,"hadoop"->3).map(x=>x._2)
res25: scala.collection.immutable.Iterable[Int] = List(1, 2, 3)

scala> Map("spark"->1,"hive"->2,"hadoop"->3).map(x=>x._1)
res26: scala.collection.immutable.Iterable[String] = List(spark, hive, hadoop)
```

（2）flatMap 函数

定义：def flatMap[B](f: (A) ⇒ GenTraversableOnce[B]): Array[B]

用途：将函数 f 作用于集合中的所有元素，各元素得到相应的集合 GenTraversableOnce[B]，然后再将其扁平化返回，生成新的集合。

```
scala> val listInt=List(1,2,3)
listInt: List[Int] = List(1, 2, 3)

//函数作用于各元素时，要求返回的是集合
scala> listInt.flatMap(x => x match {
  case 1 => List(1)
  case _ => x
})
<console>:11: error: type mismatch;
 found   : Int
 required: scala.collection.GenTraversableOnce[?]
            case _ => x
                      ^
//正确写法
scala> listInt.flatMap(x => x match {
  case 1 => List(1)
  case _ => List(x)
})
res1: List[Int] = List(1, 2, 3)
```

了解函数 f 作用于各元素之后返回的结果类型后,现在可以来看看 map 函数与 flatmap 函数之间的区别。

```
scala> listInt.map(x => x match {
  case 1 => List(1)
  case _ => List(x*2,x*3,x*4)
})
res6: List[List[Int]] = List(List(1), List(4, 6, 8), List(6, 9, 12))

scala> listInt.flatMap(x => x match {
  case 1 => List(1)
  case _ => List(x*2,x*3,x*4)
})
 res7: List[Int] = List(1, 4, 6, 8, 6, 9, 12)
```

对于 Array 等其他集合类型的 flatMap 函数,其作用原理是类似的,这里不再一一赘述。

(3) filter 函数

定义:def filter(p: (T) ⇒ Boolean): Array[T]

用途:返回所有满足条件 p 的元素集合。

```
scala> val listInt=List(1, 4, 6, 8, 6, 9, 12)
listInt: List[Int] = List(1, 4, 6, 8, 6, 9, 12)

//返回所有元素值大于 6 的元素构成的集合
scala> listInt.filter(x=>x>6)
res8: List[Int] = List(8, 9, 12)

//简化的写法
scala> listInt.filter(_>6)
res9: List[Int] = List(8, 9, 12)
```

Array 等类型的集合中的 filter 方法使用也类似,这里不再一一赘述。

(4) reduce 函数

定义:def reduce[A1 >: A](op: (A1, A1) ⇒ A1): A1

用途:使用函数 op 作用于集合之上,返回的结果类型为 A1。op 为特定的联合二元算子 (associative binary operator),A1 为 A 的超类。

```
//reduce 函数作用演示
scala> Array(1,2,4,3,5).reduce((x:Int,y:Int)=>{println(x,y);x+y})
(1,2)
(3,4)
(7,3)
(10,5)
res60: Int = 15
```

```
//简化的写法
scala> Array(1,2,4,3,5).reduce(_+_)
res61: Int = 15
```

reduce 函数还有两个函数变种，分别是 reduceLeft 和 reduceRight，reduceLeft 函数指的是 op 函数按集合中元素的顺序从左到右对其进行 reduce 操作，而 reduceRight 函数作用顺序则相反。

```
//reduceLeft 函数作用演示
scala> Array(1,2,4,3,5).reduceLeft((x:Int,y:Int)=>{println(x,y);x+y})
(1,2)
(3,4)
(7,3)
(10,5)
res61: Int = 15
```

```
//reduceRight 函数作用演示
scala> Array(1,2,4,3,5).reduceRight((x:Int,y:Int)=>{println(x,y);x+y})
(3,5)
(4,8)
(2,12)
(1,14)
res62: Int = 15
```

（5）fold 函数

定义：def fold[A1 >: A](z: A1)(op: (A1, A1) ⇒ A1): A1

用途：使用联合二元操作算子对集合进行 fold 操作，z 为给定的初始值。

```
//fold 函数执行过程演示
scala> Array(1,2,4,3,5).fold(0)((x:Int,y:Int)=>{println(x,y);x+y})
(0,1)
(1,2)
(3,4)
(7,3)
(10,5)
res10: Int = 15
```

```
//简化的写法
scala> Array(1,2,4,3,5).fold(0)(_+_)
res13: Int = 15
```

fold 函数也有两种函数变种：foldLeft 与 foldRight 函数，foldLeft 按从左到右的顺序进行 fold 操作，而 foldRight 则相反。

```
scala> Array(1,2,4,3,5).foldRight(0)((x:Int,y:Int)=>{println(x,y);x+y})
```

```
(5,0)
(3,5)
(4,8)
(2,12)
(1,14)
res11: Int = 15

scala> Array(1,2,4,3,5).foldLeft(0)((x:Int,y:Int)=>{println(x,y);x+y})

(0,1)
(1,2)
(3,4)
(7,3)
(10,5)
res12: Int = 15
```

5.4 闭 包

John D. Cook 给对象和闭包（Closure）下过一个经典的定义："An object is data with functions. A closure is a function with data"[1]，可以看到，闭包是由函数和运行时的数据决定的。事实上，闭包可以理解为函数和上下文，例如：

```
scala> var i=15
i: Int = 15

//定义一个函数字面量 f，函数中使用了前面定义的变量 i
scala> val f=(x:Int)=>x+i
f: Int => Int = <function1>

//执行函数
scala> f(10)
res0: Int = 25

//变量重新赋值
scala> i=20
i: Int = 20

//执行函数
scala> f(10)
res1: Int = 30
```

---

1 http://adv-r.had.co.nz/Functional-programming.html#closures

代码 val f=(x:Int)=>x+i 定义了一个函数字面量，函数中使用了自由变量 i，变量 i 在程序的运行过程中会发生变化，在函数执行时如调用 f(10)时会根据运行时变量 i 的值的不同，得到不同的运行结果。自由变量 i 在运行过程中会不断地发生变化，它处于一种开放状态，而当函数执行时，自由变量 i 的值已经被确定下来，此时可以认为在运行时它暂时处于封闭状态，这种存在从开放到封闭过程的函数被称为闭包。函数字面量 val f=(x:Int)=>x+i 中便是函数（f）+上下文（自由变量 i）的结合。

val f=(x:Int)=>x+i 只是函数闭包的一种形式，高阶函数也可以理解为一种闭包，例如：

```
//高阶函数 a，函数有两个参数：函数类型 Double=>Double 的参数 f、Double=>Unit 的参数 p
scala> def a(f:Double=>Double,p:Double=>Unit)={
  val x=f(10)
  p(x)
}
a: (f: Double => Double, p: Double => Unit)Unit

//定义一个输入参数为 Double，返回值类型为 Double 的函数
scala> val f=(x:Double)=>x*2
f: Double => Double = <function1>

//定义一个输入参数为 Double，返回值类型为 Unit 的函数
scala> val p=(x:Double)=>println(x)
p: Double => Unit = <function1>

//将定义的函数 f、p 作为高阶函数 a 的参数
scala> a(f,p)
20.0

//定义另外一个输入参数为 Double，返回值类型为 Double 的函数
scala> val f2=(x:Double)=>x*x
f2: Double => Double = <function1>

//将定义的函数 f2、p 作为高阶函数 a 的参数
scala> a(f2,p)
100.0
```

下列代码

```
def a(f:Double=>Double,p:Double=>Unit)={
  val x=f(10)
  p(x)
}
```

定义了高阶函数 a，该函数使用两个不同函数类型作为参数，f:Double=>Double 表示函数 f 的参数类型为 Double，返回值类型为 Double，p:Double=>Unit 表示函数 p 的参数类型为 Double，返回值类型为 Unit。函数在运行时会根据传入的不同函数，得到不同的运行结果，

如执行 a(f,p)和 a(f2,p)得到的结果是不一样的，它也是由函数（高阶函数 a）和上下文（传入的函数 f、f2、p）构成的。

5.5 函数柯里化

在介绍高阶函数时，我们通过下列代码定义了一个返回值为函数类型的高阶函数。

```
scala> def higherOrderFunction(factor:Int)=(x:Double)=>factor*x
higherOrderFunction: (factor: Int)Double => Double
```

高阶函数 higherOrderFunction 可以通过下列代码使用：

```
scala> higherOrderFunction(10)(50)
res14: Double = 500.0
```

higherOrderFunction(10)(50)中的 higherOrderFunction(10)返回的函数对象的类型是 Double => Double，然后再调用该函数对象，得到最终的结果，即上面的代码相当于下列两行代码：

```
scala> val f=higherOrderFunction(10)
f: Double => Double = <function1>

scala> f(50)
res16: Double = 500.0
```

higherOrderFunction(10)(50)这种函数调用方式与柯里化的函数调用方式十分相似。那什么是柯里化的函数呢？将函数 def higherOrderFunction(factor:Int)=(x:Double)=>factor*x 定义成如下形式：

```
// multiply 函数为柯里化的函数
scala> def multiply(factor:Int)(x:Double)=factor*x
multiply: (factor: Int)(x: Double)Double
```

def multiply(factor:Int)(x:Double)定义了一个柯里化的函数，柯里化的函数具有如图 5-3 所示的定义形式。

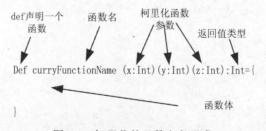

图 5-3　柯里化的函数定义形式

柯里化函数的使用方式如下：

```
//柯里化函数的使用
```

```
scala> multiply(10)(50)
res17: Double = 500.0

//柯里化函数不是高阶函数,不能像higherOrderFunction(10)这样使用返回一个函数对象
scala> multiply(10)
<console>:9: error: missing arguments for method multiply;
follow this method with `_' if you want to treat it as a partially applied f
unction
              multiply(10)
              ^
```

代码 multiply(10)会编译报错,这是因为柯里化的函数并不是高阶函数,它不会像 higherOrderFunction(10)那样返回一个函数类型的对象。正如代码出错提示 follow this method with `_' if you want to treat it as a partially applied function 那样,如果希望柯里化的函数返回生成新的函数,则需要用到下一节将介绍的部分应用函数(partially applied function)。

5.6 部分应用函数

在上一小节中,柯里化函数 def multiply(factor:Int)(x:Double)=factor*x 不能像高阶函数 def higherOrderFunction(factor:Int)=(x:Double)=>factor*x 那样,通过 multiply(10)生成新的函数,这是因为柯里化函数并不具有能够返回函数对象的高阶函数那样的性质。如果想通过 multiply(10)这样的方式生成新的函数,可以使用部分应用函数,其使用语法如下:

```
// multiply(10) _生成部分应用函数paf
scala> val paf=multiply(10) _
paf: Double => Double = <function1>

//调用部分应用函数paf
scala> paf(50)
res19: Double = 500.0
```

代码 multiply(10) _生成了一个部分应用函数,其函数类型是 Double => Double,即返回的函数要求输入参数类型是 Double,返回值类型也是 Double。事实上,柯里化的函数能够生成与其参数个数相当的部分应用函数,函数 def multiply(factor:Int)(x:Double)=factor*x 有两个参数,分别是 factor:Int 和 x:Double,它除了通过 multiply(10) _生成一个函数类型为 Double => Double 之外,还可以生成下面的部分应用函数。

```
//两个参数都不指定时对应的部分应用函数,生成的函数类型为 Int => (Double => Double)
scala> def paf2=multiply _
paf2: Int => (Double => Double)

//现在可以像higherOrderFunction那样使用
scala> paf2(10)(50)
```

```
res20: Double = 500.0

//像 higherOrderFunction 那样生成新的函数对象
scala> paf2(10)
res21: Double => Double = <function1>
```

代码 def paf2=multiply _会生成 multiply 的另外一个部分应用函数，该函数的类型是 Int => (Double => Double)即函数输入参数是 Int 类型，返回值类型是 Double=>Double（函数类型），正如代码 paf2(10)生成的结果那样，它返回的是一个函数对象 res21: Double => Double。res21 要求输入参数类型是 Double，返回值类型也是 Double。代码 paf2(10)(50)相当于先通过 paf2(10)生成一个函数对象 res21: Double => Double，再调用 res21(50)得到最终结果。

不只柯里化函数有部分应用函数，普通的函数也是如此。例如：

```
//普通函数 product
scala> def product(x1:Int,x2:Int,x3:Int)=x1*x2*x3
product: (x1: Int, x2: Int, x3: Int)Int

//一个参数的部分应用函数 product_1
scala> def product_1=product(_:Int,2,3)
product_1: Int => Int

//调用该部分应用函数
scala> product_1(2)
res22: Int = 12

//两个参数的部分应用函数 product_2
scala> def product_2=product(_:Int,_:Int,3)
product_2: (Int, Int) => Int

//调用该部分应用函数
scala> product_2(2,2)
res23: Int = 12

//三个参数的部分应用函数
scala> def product_3=product(_:Int,_:Int,_:Int)
product_3: (Int, Int, Int) => Int

//使用该部分应用函数
scala> product_3(2,2,3)
res24: Int = 12

//三个参数的部分应用函数，等价于 product(_:Int,_:Int,_:Int)
scala> def product_3=product _
```

```
product_3: (Int, Int, Int) => Int

//使用该部分应用函数
scala> product_3(2,2,3)
res24: Int = 12
```

5.7 偏函数

偏函数（Partial Function）在数学上的定义形式如图 5-4 所示，变量域中的 x2、x3、x5 及 x6 与值域中的 y2、y3、y5 及 y6 对应，但 x1、x4 在值域中找不到对应，此时称函数 f 为偏函数。

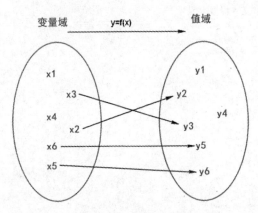

图 5-4 偏函数的数学定义

Scala 中的偏函数定义为 trait PartialFunction[-A, +B] extends (A => B)，泛型参数 A 为偏函数的输入参数类型，泛型参数 B 为偏函数的返回结果类型，由于偏函数可以只处理参数定义域的子集，对子集之外的参数会抛出异常，这一特点使得偏函数与 Scala 模式匹配能够完美结合。

```
scala> val sample = 1 to 10
sample: scala.collection.immutable.Range.Inclusive = Range(1, 2, 3, 4, 5, 6, 7, 8, 9, 10)

//定义一个偏函数 isEven，只处理为偶数值的参数
scala> val isEven: PartialFunction[Int, String] = {
  case x if x % 2 == 0 => x+" is even"
}
isEven: PartialFunction[Int,String] = <function1>

//只能处理输入的偶数值
scala> isEven(10)
res0: String = 10 is even
```

```
//处理非偶数值会报错
scala> isEven(11)
scala.MatchError: 11 (of class java.lang.Integer)
    at scala.PartialFunction$$anon$1.apply(PartialFunction.scala:248)
    at scala.PartialFunction$$anon$1.apply(PartialFunction.scala:246)
//其他出借代码省略

//返回所有在偏函数isEven作用域中的元素,执行过程中会调用def isDefinedAt(x: A): Boolean函数检查是否在作用域中
scala> sample.collect(isEven)
res3: scala.collection.immutable.IndexedSeq[String] = Vector(2 is even, 4 is even, 6 is even, 8 is even, 10 is even)

//偏函数,只处理值为奇数的参数
scala> val isOdd: PartialFunction[Int, String] = {
  case x if x % 2 == 1 => x+" is odd"
}
isOdd: PartialFunction[Int,String] = <function1>

scala> isOdd(11)
res4: String = 11 is odd

// def orElse[A1 <: A, B1 >: B](that: PartialFunction[A1, B1]): PartialFunction[A1, B1]方法将参数作用域合并
scala> val numbers = sample.map(isEven orElse isOdd)
numbers: scala.collection.immutable.IndexedSeq[String] = Vector(1 is odd, 2 is even, 3 is odd, 4 is even, 5 is odd, 6 is even, 7 is odd, 8 is even, 9 is odd, 10 is even)
```

上面的代码演示的是将变量定义为 PartialFunction[Int,String]类型,在实际中还可以在定义函数时使用,例如:

```
//定义函数时使用PartialFunction[Any, Unit]修饰
scala> def receive:PartialFunction[Any, Unit]={
  case x:Int=>println("Int Type")
  case x:String=>println("String Type")
  case _=>println("Other Type")
}
receive: PartialFunction[Any,Unit]

scala> receive(10)
Int Type
```

```
scala> receive("10")
String Type
```

代码 def receive:PartialFunction[Any, Unit]在定义函数 receive 时使用 PartialFunction[Any, Unit]修饰，表示函数 receive 可以接收任意类型的参数且函数没有返回值，事实上 def receive:PartialFunction[Any, Unit]函数定义方式等同于 val receive:PartialFunction[Any, Unit]定义方式，具体验证代码如下：

```
scala> val receive:PartialFunction[Any, Unit]={
    case x:Int=>println("Int Type")
    case x:String=>println("String Type")
    case _=>println("Other Type")
}
receive: PartialFunction[Any,Unit] = <function1>

scala> receive(10)
Int Type

scala> receive("10")
String Type
```

小　结

本章介绍了 Scala 函数式编程，内容涵盖函数、函数字面量、高阶函数、函数闭包、函数柯里化、部分应用函数及偏函数的定义与使用。重点介绍了函数的简化、高阶函数如何接受函数类型参数及返回函数类型，柯里化函数的表现形式及部分应用函数的使用。在下一章中，我们将会介绍 Scala 面向对象编程的相关内容。

第6章 Scala 面向对象编程（上）

  Scala 是一门纯面向对象编程语言，Scala 中的一切变量都是对象，一切操作都是方法调用。在本章中将对 Scala 中类的定义、对象的创建、主构造函数的定义与使用、辅助构造函数的定义与使用、继承、类成员访问控制、抽象类等内容进行深入介绍。

6.1 类与对象

6.1.1 类的定义

Scala 中的类与 Java 语言中的类一样都是通过 class 关键字来定义。

```scala
//class 关键字声明一个类 Person
class Person {
  //声明一个成员变量，这里成员变量必须初始化
  var name:String=null
}
```

将 Person 生成的字节码文件进行反编译，得到如下内容：

```
E:\IntellijIDEAWorkspace\out\production\ScalaProject\cn\scala\chapter06>javap -private Person.class
Compiled from "Person.scala"
public class cn.scala.chapter06.Person {
  private java.lang.String name;
  public java.lang.String name();
  public void name_$eq(java.lang.String);
  public cn.scala.chapter06.Person();
}
```

通过上述代码可以看到，Person 类会自动生成 3 个方法，它们分别是 public java.lang.String name()、public void name_$eq(java.lang.String) 及默认构造函数 public cn.scala.chapter06.Person()。成员方法 name() 为 Scala 风格的 getter 方法，而方法 name_$eq(java.lang.String) 为 Scala 风格的 setter 方法，name_$eq(java.lang.String) 方法其实等同于 name_=(java.lang.String)，这是由于字节码文件的符号限制导致的。值得注意的是，在定义 Person 类时，成员变量 var name:String=null 前面没有加任何关键字，但通过反编译后的字节码文件可以看到，成员变量 name 为 private，因此访问成员变量 name 必须通过 name() 及 name_$eq 方法。

定义 Person 类时还有个值得注意的地方，就是类的成员变量在声明时必须初始化，否则会报错。

```
scala> class Person{
  var name:String
}
<console>:7: error: class Person needs to be abstract, since variable name is not defined
(Note that variables need to be initialized to be defined)
       class Person{
       ^
```

成员变量如果在定义时确实不需要初始化，则需要将类定义为抽象类，关于抽象类将会在 6.6 节中详细介绍。成员变量初始化时还可以使用占位符的方式，例如：

```
scala> class Person{
  var name:String=_
}
defined class Person
```

使用占位符 "_" 对成员变量初始化时，如果是值对象类型如 Double、Int 等，在 Scala 中对应 AnyVal 类，则会初始化为 0，如果是引用类型如 String 等，在 Scala 中对应 AnyRef 类，则会初始化为 Null。

6.1.2 创建对象

在 Scala 语言中，创建对象同样通过 new 关键字进行。

```
//通过 new 关键字创建 Person 类的对象
scala> val p=new Person()
p: Person = Person@1d48f21


//Person 类的构造函数无参数，可以省略()
scala> val p1=new Person
p1: Person = Person@d22908
```

6.1.3 类成员的访问

创建 Person 类对象后可以通过方法 name()和 name_=(java.lang.String)访问成员变量 name。

```
//显式调用 setter 方法修改成员变量 name
scala> p.name_=("john")

//调用 getter 方法访问成员变量 name
scala> p.name
res2: String = john

//直接修改成员变量 name，但实际上调用的是 p.name_方法
scala> p.name="John"
p.name: String = John

scala> p.name
res3: String = John
```

代码中可以显式地调用 setter 方法 p.name_=("john")对成员变量 name 进行修改，也可以通过 p.name="John"直接修改成员变量，但通过前面的 Person 字节码反编译后的代码来看，成

员变量 name 是私有成员，并不能直接访问该私有成员，代码 p.name="John" 之所以合法是因为它其实调用的仍然是 setter 方法 name_=(java.lang.String)，只不过编译器为开发人员屏蔽了相应的细节。像这种调用者并不知道是直接对成员变量进行访问的，还是通过 setter 方法对成员变量进行访问的方式被称为统一访问原则。

前面在 Person 类的成员变量被声明为 var，在使用时成员变量的值可以通过成员方法对其进行修改，如果成员变量一旦被赋值便不能被修改，此时可以将其声明为 val，代码如下：

```scala
class Person {
    val name:String=null
}
```

将生成的字节码文件反编译后的内容如下：

```
E:\IntellijIDEAWorkspace\out\production\ScalaProject\cn\scala\chapter06>javap -private Person.class
Compiled from "Person.scala"
public class cn.scala.chapter06.Person {
  private final java.lang.String name;
  public java.lang.String name();
  public cn.scala.chapter06.Person();
}
```

可以看到，成员变量 name 被声明为 val 之后，只会生成 getter 方法 public java.lang.String name()，不会生成 setter 方法 name_=(java.lang.String)，使用时不能对成员变量进行修改，如下所示：

```
scala> p.name
res0: String = null

scala> p.name="John"
<console>:9: error: reassignment to val
    p.name="John"
           ^
```

Scala 风格的 getter 与 setter 方法提供了统一访问成员变量的形式，但在实际开发应用中可能会碰到需要生成 Java 风格的 getter 与 setter 方法的情况，此时可以通过对成员变量进行 @BeanProperty 注解来实现，示例如下：

```scala
import scala.beans.BeanProperty
class Person {
    @BeanProperty var name:String=null
}
```

对上述代码编译后的字节码文件进行反编译，则有：

```
E:\IntellijIDEAWorkspace\out\production\ScalaProject\cn\scala\chapter06>javap -private Person.class
```

```
Compiled from "Person.scala"
public class cn.scala.chapter06.Person {
  private java.lang.String name;
  public java.lang.String name();
  public void name_$eq(java.lang.String);
  public void setName(java.lang.String);
  public java.lang.String getName();
  public cn.scala.chapter06.Person();
}
```

可以看到，成员变量前面加上@BeanProperty 后，除了生成 Scala 风格的 getter 与 setter 方法之外，还会生成像 public void setName(java.lang.String)、public java.lang.String getName() 这种 Java 风格的 getter 与 setter 方法，此时便可以使用 setName 及 getName 方法直接访问类的成员变量，例如：

```
scala> p.setName("John")
scala> p.getName
res1: String = John
```

6.1.4 单例对象

在实际应用中，常常存在不创建对象就可以访问相应的成员变量或方法的场景，Java 语言中提供了静态成员和静态方法来支持这一场景。例如：

```
class Utils{
    public static Double PI=3.141529;
    public static String getName(){
        return "圆周率";
    }
}
public class JavaExample6_01 {
    public static void main(String[] args) {
        System.out.println(Utils.PI);
        System.out.println(Utils.getName());
    }
}
```

在类 Utils 当中，通过 static 关键字来声明静态类成员 PI 和 getName，使用时可以不创建对象而是直接通过类名.类成员名称的方式来访问。

Scala 语言并不支持静态类成员这一语法，而是通过单例对象来实现，如下面的示例：

```
//通过 object 关键字声明单例对象 Student
scala> object Student {
  private var studentNo:Int=0;
  def uniqueStudentNo()={
    studentNo+=1
```

```
       studentNo
   }
}
defined module Student

//直接通过单例对象名称访问其成员方法
scala> Student.uniqueStudentNo()
res1: Int = 1

//单例对象中的成员变量状态随程序执行而改变
scala> Student.uniqueStudentNo()
res2: Int = 2
```

查看单例对象编译后生成的字节码文件，可以发现它会生产两个字节码文件：Student$.class 和 Student.class，对 Student$.class 字节码文件进行反编译，得到如下代码：

```
E:\IntellijIDEAWorkspace\out\production\ScalaProject\cn\scala\chapter06>javap -private Student$.class
Compiled from "Person.scala"
public final class cn.scala.chapter06.Student$ {
  public static final cn.scala.chapter06.Student$ MODULE$;
  private int studentNo;
  public static {};
  private int studentNo();
  private void studentNo_$eq(int);
  public int uniqueStudentNo();
  private cn.scala.chapter06.Student$();
}
```

而对 Student.class 字节码文件进行反编译，得到如下代码：

```
E:\IntellijIDEAWorkspace\out\production\ScalaProject\cn\scala\chapter06>javap -private Student.class
Compiled from "Person.scala"
public final class cn.scala.chapter06.Student {
  public static int uniqueStudentNo();
}
```

通过代码不难发现，cn.scala.chapter06.Student$ 中使用了单例模式，而类 cn.scala.chapter06.Student 中有一个静态方法，也就是 Scala 单例对象实际上通过 Java 语言的单例模式和静态类成员来实现。

6.1.5 应用程序对象

在 Java 语言中，只要类中定义了 public static void main(String[] args)方法，该类便可以成为程序执行的入口。在 Scala 语言中同样使用 main 函数作为程序的执行入口，只不过 main

函数必须定义在单例对象中。例如：

```scala
/**
 * 应用程序对象：通过定义main方法来实现
 */
object Example6_01{
  object Student {
    private var studentNo:Int=0;
    def uniqueStudentNo()={
      studentNo+=1
      studentNo
    }
  }
  //通过main方法作为程序的入口
  def main(args: Array[String]) {
    println(Student.uniqueStudentNo())
  }
}
```

在单例对象 Example6_01 中，定义了一个 main 方法 def main(args: Array[String])，该 main 方法将作为程序的入口。需要注意的是，在上面的代码中将单例对象 Student 定义在单例对象 Example6_01 中，这在 Scala 语言中是合法的，在后续章节中我们还会大量地使用这种方式。只要单例对象中定义了 main 方法，该单例对象便被称为应用程序对象。

应用程序对象还可以通过混入 trait App 来定义，示例如下：

```scala
/**
 * 应用程序对象：通过混入trait App 来实现
 */
object Example6_01 extends App{
  object Student {
    private var studentNo:Int=0;
    def uniqueStudentNo()={
      studentNo+=1
      studentNo
    }
  }
  //直接写可执行的代码
  println(Student.uniqueStudentNo())
}
```

可以看到，通过 extends App 来定义应用程序对象，能够使代码更简洁，直接在单例对象中编写 println(Student.uniqueStudentNo())这样的可执行代码即可，无需将代码放在自定义的 main 函数中。事实上，通过 extends App 定义应用程序对象的方式仍然是使用 main 函数作为程序的入口，查看 trait App 的源代码可以看到，App 中已经为我们定义了 main 函数，源代码如下：

```
def main(args: Array[String]) = {
  this._args = args
  for (proc <- initCode) proc()
  if (util.Properties.propIsSet("scala.time")) {
    val total = currentTime - executionStart
    Console.println("[total " + total + "ms]")
  }
}
```

实际中最常见的应用程序对象还是通过 extends App 的方式来定义，因为这种方式会使代码更简洁。

6.1.6 伴生对象与伴生类

在 Scala 中还存在伴生对象（Companion Object）和伴生类（Companion Class）的概念，在前一小节中定义了一个单例对象 object Student，为说明伴生对象与伴生类的作用，现在我们再定义一个类名为 Student 的类。请看下列代码：

```
/**
 * 伴生对象与伴生类
 */
object Example6_02 extends App{
  //伴生类 Student
  class Student{
    private var name:String=null
  }

  //伴生对象
  object Student {
    private var studentNo:Int=0;
    def uniqueStudentNo()={
      studentNo+=1
      studentNo
    }
  }
}
```

在 Example6_02 中，有一个单例对象 Student 和类 Student，此时的 object Student 称为 class Student 的伴生对象，而 class Student 则称为 object Student 的伴生类。

伴生类与伴生对象在 Scala 中是十分重要的两个概念，伴生类与伴生对象区别于其他类或对象十分重要的地方便是访问控制权限，在本章 6.5 节中的成员访问控制部分我们会进行更详细的介绍，这里先简单演示伴生类与伴生对象间的成员访问，先看下面的例子。

```
/**
 * 伴生对象与伴生类
```

```scala
   */
  object Example6_02 extends  App{
    //伴生类Student
    class Student{
      private var name:String=null
      def getStudentNo={
        Student.uniqueStudentNo()
        //伴生类Student中可以直接访问伴生对象Student的私有成员
        Student.studentNo
      }
    }

    //伴生对象
    object Student {
      private var studentNo:Int=0;
      def uniqueStudentNo()={
        studentNo+=1
        studentNo
      }
      //直接访问伴生类对象的私有成员
      def printStudenName=println(new Student().name)
    }

    //在伴生类和伴生对象外部不允许直接访问伴生对象Student的私有成员studentNo
    //println(Student.studentNo)

    //在伴生类和伴生对象外部不允许访问伴生类Student对象的私有成员name
    //println(new Student().name)

    println(new Student().getStudentNo)
    println(Student.printStudenName)
  }
```

上面的代码给出了伴生对象与伴生类的使用代码，可以看到在伴生类 Student 的 getStudentNo 方法中，可以通过 Student.studentNo 直接访问伴生对象的私有成员，但在伴生类和伴生对象外部不能这样使用，也就是说试图在外部通过 println(Student.studentNo)来访问伴生对象的私有成员 studentNo 是不被允许的。同样在伴生对象 Student 中的成员方法 def printStudenName=println(new Student().name)中可以直接访问伴生类对象的私有成员变量 name，而在伴生类和伴生对象外部，是不被允许的，代码 println(new Student().name)说明了这一点。通过该例子可以发现，伴生类与伴生对象开放了对彼此成员的访问权限，除了这个便利之外，伴生对象与伴生类还可以实现无 new 创建对象及析构对象。关于使用伴生类和伴生对象实现对象析构这部分内容将会放在第 9 章模式匹配中进行介绍，本节先介绍如何通过伴生类和伴生对象实现无 new 方式创建对象。

通过非显式使用 new 关键字创建对象，在集合那部分已经大量使用，如下述代码：

```
scala> Array(1,2,3,4,5)
res3: Array[Int] = Array(1, 2, 3, 4, 5)
```

其实它调用的是伴生对象 Array 中的 apply 方法。

```
/** Creates an array of `Int` objects */
// Subject to a compiler optimization in Cleanup, see above.
def apply(x: Int, xs: Int*): Array[Int] = {
  val array = new Array[Int](xs.length + 1)
  array(0) = x
  var i = 1
  for (x <- xs.iterator) { array(i) = x; i += 1 }
  array
}
```

如果希望在编写代码时，不通过 new Student()这样显式地调用 new 关键字来创建对象，可以使用如下方式：

```
/**
 * 伴生对象与伴生类：不显式地通过 new 关键字创建对象
 */
object Example6_03 extends App{
  //伴生类 Student
  class Student{
    var name:String=null
  }

  //伴生对象
  object Student {
    //apply 方法，供非显式 new 创建对象时调用
    def apply()=new Student()
  }

//无 new 方式创建对象，调用的是伴生对象的 apply 方法：def apply()=new Student()
  val s=Student()
  s.name="John"
  //打印输出 John
  println(s.name)
}
```

在伴生对象 object Student 中定义了一个 apply 方法：def apply()=new Student()，在执行代码 val s=Student()时会自动调用伴生对象 Student 的 apply 方法，完成伴生类 Student 对象的创建。

6.2 主构造函数

6.2.1 主构造函数的定义

在前面的例子中，在添加成员变量时仍然沿用的是 Java 风格的成员变量定义方式，例如：

```
class Student{
    var name:String=null
}
```

Scala 会自动生成一个默认的构造函数和 Scala 风格的 getter 方法 public java.lang.String name()和 setter 方法 public void name_=(java.lang.String)。事实上 Scala 提供了更简化的成员变量和构造函数定义方式，即将成员变量和构造函数交织在一起的定义方式，例如：

```
//带主构造函数的类定义
class Person(var name:String,var age:Int)
```

为说明这种类定义背后的原理，先对其生成的字节码文件进行反编译，得到下述代码：

```
E:\IntellijIDEAWorkspace\out\production\ScalaProject\cn\scala\chapter06>javap -private Example6_04$Person.class
Compiled from "Example6_01.scala"
public class cn.scala.chapter06.Example6_04$Person {
  private java.lang.String name;
  private int age;
  public java.lang.String name();
  public void name_$eq(java.lang.String);
  public int age();
  public void age_$eq(int);
  public cn.scala.chapter06.Example6_04$Person(java.lang.String, int);
}
```

可以看到，代码 class Person(var name:String,var age:Int)做了 3 件事情：

- 定义了一个类 Person 并实现了构造函数 Person(java.lang.String, int)。
- 定义了 name 和 age 两个成员变量。
- 分别生成了 name 和 age 对应 Scala 风格的 setter 方法和 getter 方法。

如果使用 Java 语言来实现代码 class Person(var name:String,var age:Int)同样功能的话，则需要使用下列代码：

```
public class Person{
    private  String name;
    private  int age;
    public String getName() {
        return name;
```

```java
    }
    public void setName(String name) {
        this.name = name;
    }
    public int getAge() {
        return age;
    }
    public void setAge(int age) {
        this.age = age;
    }
    public Person(String name, int age) {
        this.name = name;
        this.age = age;
    }
}
```

可以看到，Scala 只使用一行代码就可以达到 Java 多行代码才能实现的功能。

在 Java 语言中，如果希望在创建对象时执行某些动作，则需要在构造函数中添加相应的代码，但在 Scala 中只需要将代码直接放在定义的类中，例如：

```scala
//当在创建对象时，需要进行相关初始化操作时，可以将初始化语句放在类体中
class Person(val name:String,val age:Int){
  //println 将作为主构建器中的一部分，在创建对象时被执行
  println("constructing Person ........")
  //重写 toString()方法
  override def toString()= name + ":"+ age
}


//创建对象时将自动执行类中的程序语句
scala> val p=new Person("john",29)
constructing Person ........
p: Person = john:29
```

在上面的代码中，在 Person 类体中放置了一个执行语句 println("constructing Person ....")，这条语句便会作为主构造函数的一部分，在创建对象时被执行。它相当于下列 Java 代码：

```java
public class Person{
    private  String name;
    private  int age;

    public String getName() {
        return name;
    }
    public int getAge() {
```

```java
        return age;
    }

    public Person(String name, int age) {
        System.out.println("constructing Person ........");
        this.name = name;
        this.age = age;
    }
}
```

6.2.2 默认参数的主构造函数

主构造函数还可以带默认参数，这样在创建对象时可以不指定参数而使用默认值，带默认参数的主构造函数定义如下：

```scala
//默认参数的主构造函数
class Person(var name:String="",var age:Int=18)
```

代码指定成员变量 name 的默认值为""，而成员变量 age 的默认值为 18，这样在创建对象时可以不指定参数，具体示例代码如下：

```scala
//默认参数的主构造函数
scala> class Person(var name:String="",var age:Int=18){
     override def toString()="name="+name+",age="+age
}
defined class Person

//不指定参数时，使用默认值
scala> val p=new Person
p: Person = name=,age=18

//指定 name，但不指定 age
scala> val p1=new Person("John")
p1: Person = name=John,age=18
```

主构造函数中默认参数也可只指定部分，例如：

```scala
//指定部分参数为默认参数
scala>class Person(var name:String,var age:Int=18){
  override def toString()="name="+name+",age="+age
}
defined class Person

scala> val p=new Person
<console>:8: error: not enough arguments for constructor Person: (name: Str
```

```
ing, age: Int)Person:
   Unspecified value parameter name.
        val p=new Person
         ^

//给成员变量 name 指定非默认参数值，而成员变量 age 则使用默认参数值
scala> val p=new Person("Jchn")
p: Person = name=John,age=18
```

6.2.3 私有主构造函数

上节代码 class Person(var name:String,var age:Int)中的主构造函数是公有的，在创建对象时可以直接调用，示例如下：

```
//通过主构造函数创建对象
scala> val p=new Person("John",18)
p: Person = Person@f820d8
```

但在实际应用中，可能希望将主构造函数私有化，只在类内部使用而不对外开放，此时可以将主构造函数通过关键字 private 定义为私有。

```
//通过 private 关键字将主构造函数定义为私有
scala> class Person private(var name:String,var age:Int)
defined class Person

//只能在类内部使用，外部不能使用
scala> val p=new Person("John",19)
<console>:8: error: constructor Person in class Person cannot be accessed in object $iw
      val p=new Person("John",19)
```

对 class Person private(var name:String,var age:Int)生成的字节码文件进行反编译，得到如下代码：

```
E:\IntellijIDEAWorkspace\out\production\ScalaProject\cn\scala\chapter06>javap -private Example6_04$Person.class
Compiled from "Example6_01.scala"
public class cn.scala.chapter06.Example6_04$Person {
  private final java.lang.String name;
  private final int age;
  public java.lang.String name();
  public int age();
  private cn.scala.chapter06.Example6_04$Person(java.lang.String, int);
}
```

可以看到，此时的构造函数 Person(java.lang.String, int)已经是 private。

6.3 辅助构造函数

6.3.1 辅助构造函数的定义

Java 语言中在定义类的多个构造函数时，构造函数名称与类名必须相同，例如：

```java
//Java 风格的构造函数，构造函数名称与类名一致
public Person(String name) {
    this.name = name;
}

public Person(int age) {
    this.age = age;
}

public Person(String name, int age) {
    this.name = name;
    this.age = age;
}
```

Java 这种构造函数定义方式在实际使用时会造成不少麻烦，如修改类名时构造函数名称也必须进行修改，虽然集成开发环境可以在一定程度上解决该问题，但这并不是理想的方式。Scala 语言在定义辅助构造函数时通过使用 this 关键字来解决这一问题。例如：

```scala
/**
 * 辅助构造函数
 */
object Example6_05 extends  App{

  //定义无参的主构造函数
  class Person{
    //类成员
    private var name:String=null
    private var age:Int=18
    private var sex:Int=0

    //辅助构造函数
    def this(name:String){
      //this()调用的是无参的默认主构造函数
      this()
      this.name=name
```

```scala
  }
  def this(name:String,age:Int){
    //this(name)调用的是前面定义的辅助构造函数 def this(name:String)
    this(name)
    this.age=age
  }
  def this(name:String,age:Int,sex:Int){
    //this(name)调用的是前面定义的辅助构造函数 def this(name:String,age:Int)
    this(name,age)
    this.sex=sex
  }

  override def toString = {
    val sexStr=if(sex==1) "男" else "女"
    s"name=$name,age=$age,sex=$sexStr"
  }
}

//调用的是 3 个参数的辅助构造函数
println(new Person("John",19,1))
//调用的是 2 个参数的辅助构造函数
println(new Person("John",19))
//调用的是 1 个参数的辅助构造函数
println(new Person("John"))
}
```

程序运行结果如下：

```
name=John,age=19,sex=男
name=John,age=19,sex=女
name=John,age=18,sex=女
```

代码中 class Person 后面没有参数，这其实是定义了一个无参的主构造函数，在类中分别通过 def this(name:String)、def this(name:String,age:Int)及 def this(name:String,age:Int,sex:Int)定义了 3 个辅助函数，辅助构造函数通过 this 关键字来定义，而且在类内部使用辅助构造函数时同样也是通过 this 关键字实现调用，如 3 个参数的辅助构造函数 def this(name:String,age:Int,sex:Int)中使用 this(name,age)这种方式调用两个参数的辅助构造函数 def this(name:String,age:Int)。在使用辅助构造函数创建对象时，可直接通过类名并指定相应的参数来完成，编译器会根据参数的个数及类型决定调用哪个辅助函数。

对 Example6_05 中的 Person 类所生成的字节码文件进行反编译，得到如下代码：

```
E:\IntellijIDEAWorkspace\out\production\ScalaProject\cn\scala\chapter06>javap -private Example6_05$Person.class
Compiled from "Example6_01.scala"
public class cn.scala.chapter06.Example6_05$Person {
```

```
    private java.lang.String name;
    private int age;
    private int sex;
    private java.lang.String name();
    private void name_$eq(java.lang.String);
    private int age();
    private void age_$eq(int);
    private int sex();
    private void sex_$eq(int);
    public java.lang.String toString();
    public cn.scala.chapter06.Example6_05$Person();
    public cn.scala.chapter06.Example6_05$Person(java.lang.String);
    public cn.scala.chapter06.Example6_05$Person(java.lang.String, int);
    public cn.scala.chapter06.Example6_05$Person(java.lang.String, int, int);
}
```

可以看到，共生成了 4 个构造函数，分别是 Person()、Person(java.lang.String)、Person(java.lang.String, int)及 Person(java.lang.String, int, int)，其中 Person()为主构造函数，其他构造函数分别对应前面我们定义的不同参数的辅助构造函数。

6.3.2 辅助构造函数中的默认参数

同主构造函数一样，辅助构造函数中也可以有默认参数，示例如下：

```
/**
  * 辅助构造函数:默认参数
  */
object Example6_06 extends  App{

  //定义无参的主构造函数
  class Person{
    //类成员
    private var name:String=null
    private var age:Int=18
    private var sex:Int=0

    //带默认参数的辅助构造函数
    def this(name:String="",age:Int=18,sex:Int=1){
      //先调用主构造函数，这是必须的，否则会报错
      this()
      this.name=name
      this.age=age
      this.sex=sex
    }
```

```
    override def toString = {
      val sexStr=if(sex==1) "男" else "女"
      s"name=$name,age=$age,sex=$sexStr"
    }
  }
  //使用默认参数,调用的是辅助构造函数,相当于调用Person("John",18,1)
  println(new Person("John"))
  //注意这里调用的是无参的主构造函数,而不是带有默认参数的辅助构造函数
  println(new Person)
}
```

代码运行结果如下：

```
name=John,age=18,sex=男
name=null,age=18,sex=女
```

代码中只定义了一个辅助构造函数 def this(name:String="",age:Int=18,sex:Int=1)，该构造函数带有 3 个默认参数，在创建对象时可以只指定部分参数，其余部分参数则会使用默认值。例如执行代码 println(new Person("John"))时，new Person("John")会调用辅助构造函数并使用默认值完成对象的创建，即 new Person("John")相当于 new Person("John",18,1)。需要特别注意的是，println(new Person)中的 new Person 不会调用辅助构造函数，从执行时返回的结果"name=null,age=18,sex=女"可以看到，实际上调用的是无参数的主构造函数。不难看出，在创建对象时，编译器会首先查找是否有主构造函数与待调用的构造函数相匹配，有则调用，没有才会查找相应的辅助构造函数。

在讲主构造函数时，我们提到在某些情况下可能会将主构造函数定义为私有，此时创建对象便可以通过辅助构造函数来进行。对 Example6_06 中的代码进行修改将其主构造函数定义为私有，则使用 new Person 创建对象时会调用带默认参数的辅助函数，示例代码如下：

```
/**
 * 辅助构造函数:将主构造函数定义为private
 */
object Example6_07 extends App{

  //定义无参的主构造函数并将其定义为私有
  class Person private{
    //类成员
    private var name:String=null
    private var age:Int=18
    private var sex:Int=0

    //带默认参数的主构造函数
    def this(name:String="",age:Int=18,sex:Int=1){
      //先调用主构造函数,这是必须的,否则会报错
      this()
      this.name=name
```

```
      this.age=age
      this.sex=sex
   }

   override def toString = {
      val sexStr=if(sex==1) "男" else "女"
      s"name=$name,age=$age,sex=$sexStr"
   }
}
//使用默认参数,调用的是辅助构造函数,相当于调用Person("John",18,1)
println(new Person("John"))
//由于主构造函数为private,此时调用的便是带默认参数的辅助构造函数
//相当于调用Person("",18,1)
println(new Person)
}
```

示例运行结果如下:

```
name=John,age=18,sex=男
name=,age=18,sex=男
```

Example6_07 中通过代码 class Person private{…}将主构造函数定义为 private,此时便不能通过无参的主构造函数创建对象,从 println(new Person)的运行结果 "name=,age=18,sex=男" 可以看到,它调用的是带默认参数的辅助函数完成对象的创建,new Person 相当于 new Person("",18,1)。因为主构造函数被定义为 private,通过 new Person 创建对象时,编译器查找不到能够访问的主构造函数,从而会使用带默认参数的辅助构造函数。

6.4 继承与多态

继承是面向对象编程语言实现代码复用的关键特性,它是从一般到特殊的过程,指的是在原有类的基础上定义一个新的类,原有类称为父类,新的类称为子类。实现继承后,子类可以拥有父类的属性和方法,也可以在子类中添加新的类成员,同时又可以根据子类具体的行为重写父类的方法。多态是在继承的基础上实现的一种语言特性,它指的是允许不同类的对象对同一消息做出响应,即同一消息可以根据发送对象的不同而采用多种不同的行为方式。本小节将对 Scala 语言中的继承与多态进行详细介绍。

6.4.1 类的继承

Scala 语言同 Java 语言一样,也是通过 extends 关键字来实现类间的继承的,例如:

```
/**
 * 类继承
 */
```

```
object Example6_08 extends App{
  //定义Person类，带主构造函数
  class Person(var name:String,var age:Int){
    override def toString: String = "name="+name+",age="+age
  }

  //通过extends关键字实现类的继承，注意Student中的name和age前面没有var关键字修饰
  //表示这两个成员继承自Person类，var studentNo:String则为Student类中定义的新成员
  class Student(name:String,age:Int,var studentNo:String) extends Person(name,age){
    override def toString: String = super.toString+",studentNo="+studentNo
  }

  println(new Person("John",18))
  println(new Student("Nancy",19,"140116"))
}
```

示例运行结果如下：

```
name=John,age=18
name=Nancy,age=19,studentNo=140116
```

Example6_08 中通过代码 class Person(var name:String,var age:Int) 定义了一个带主构造函数的 Person 类，然后通过代码 class Student(name:String,age:Int,var studentNo:String) extends Person(name,age)来实现 Student 与 Person 类间的继承。

Person 类中成员变量 name、age 都被定义为 var，如果 Student 类中的 name 和 age 成员变量要从父类 Person 中继承的话，则不能使用 var 或 val 关键字修饰，extends 关键字后面的 Person(name,age)表示 Student 类继承自 Person 类且将其成员变量也一并继承。类 Student 主构造函数中的 var studentNo:String 为 Student 类中新增的成员变量。

对 Student 类和 Person 类生成的字节码文件进行反编译，得到下述代码：

```
//Person类字节码文件反编译后的代码
E:\IntellijIDEAWorkspace\out\production\ScalaProject\cn\scala\chapter06>javap -private Example6_08$Person.class
Compiled from "Example6_01.scala"
public class cn.scala.chapter06.Example6_08$Person {
  private java.lang.String name;
  private int age;
  public java.lang.String name();
  public void name_$eq(java.lang.String);
  public int age();
  public void age_$eq(int);
  public java.lang.String toString();
```

```
        public cn.scala.chapter06.Example6_08$Person(java.lang.String, int);
    }

    //Student 类字节码文件反编译后的代码
    E:\IntellijIDEAWorkspace\out\production\ScalaProject\cn\scala\chapter06>ja
vap -private Example6_08$Student.class
    Compiled from "Example6_01.scala"
    public class cn.scala.chapter06.Example6_08$Student extends cn.scala.chapt
er06.Etxample6_08$Person {
        private java.lang.String studentNo;
        public java.lang.String studentNo();
        public void studentNo_$eq(java.lang.String);
        public java.lang.String toString();
        public cn.scala.chapter06.Example6_08$Student(java.lang.String, int, java.
langString);
    }
```

通过观察反编译后的代码不难发现，Student 类继承自 Person 类并在类中定义了相应的成员变量 studentNo 及实现相应的 getter、setter 方法，同时还定义了 3 个参数的构造函数。Person 类与 Student 类如果用 Java 语言来实现的话，则其代码为：

```
class Person {
    private String name;
    private int age;
    public String getName() {
        return name;
    }
    public void setName(String name) {
        this.name = name;
    }
    public int getAge() {
        return age;
    }
    public void setAge(int age) {
        this.age = age;
    }
    public Person(String name) {
        this.name = name;
    }
    public Person(int age) {
        this.age = age;
    }
    public Person(String name, int age) {
        this.name = name;
        this.age = age;
```

```java
    }
    @Override
    public String toString() {
        return "Person{" +
                "name='" + name + '\'' +
                ", age=" + age +
                '}';
    }
}

class Student extends Person {
    private String studentNo;
    public String getStudentNo() {
        return studentNo;
    }
    public void setStudentNo(String studentNo) {
        this.studentNo = studentNo;
    }
    public Student(String name, int age, String studentNo) {
        super(name, age);
        this.studentNo = studentNo;
    }
    @Override
    public String toString() {
        return super.toString() + "Student{" +
                "studentNo='" + studentNo + '\'' +
                '}';
    }
}
```

对比 Scala 语言代码和 Java 语言代码不难发现，相同功能的程序，Scala 编写的代码要比 Java 语言编写的代码更简洁、更优雅。

6.4.2 构造函数执行顺序

在 Java 语言中，如果两个类之间存在继承关系，创建子类对象时会先调用父类的构造函数，然后再调用子类的构造函数来完成对象的创建。在 Scala 语言中，创建子类对象时的构造函数执行顺序也是如此，例如：

```scala
/**
 * 类继承：构造函数执行顺序
 */
object Example6_09 extends App{
```

```
//定义Person类,带主构造函数
class Person(var name:String,var age:Int){
    //类中的执行语句会在调用主构造函数时会执行
    println("执行Person类的主构造函数")
}

//定义Student类,继承自Person类,同样也带主构造函数
class Student(name:String,age:Int,var studentNo:String) extends Person(name,age){
    //类中的执行语句会在调用主构造函数时会执行
    println("执行Student类的主构造函数")
}

//创建子类对象时,先调用父类的主构造函数,然后再调用子类的主构造函数
new Student("Nancy",19,"140116")
```

代码执行结果如下:

执行Person类的主构造函数
执行Student类的主构造函数

通过执行结果不难看出,在使用代码new Student("Nancy",19,"140116")创建对象时,首先会调用Person类中的主构造函数,然后再调用Student类的主构造函数。

6.4.3 方法重写

方法重写指的是当子类继承父类的时候,从父类继承过来的方法不能满足子类的需要,子类希望有自己的实现,这时需要对父类的方法进行重写(override),方法重写是实现多态和动态绑定的关键。 Scala语言中的方法重写与Java语言中的方法重写一样,也是通过override关键字对父类中的方法进行重写,从而实现子类自身处理逻辑。例如:

```
/**
 * 类继承:方法重写
 */
object Example6_10 extends App{

    class Person(var name:String,var age:Int){
        //对父类Any中的toString方法进行重写
        override def toString = s"Person($name, $age)"
    }

    class Student(name:String,age:Int,var studentNo:String) extends Person(name,age){
        //对父类Person中的toString方法进行重写
```

```
    override def toString = s"Student($name,$age,$studentNo)"
  }

  //调用 Student 类自身的 toString 方法返回结果
  println(new Student("Nancy",19,"140116"))
}
```

代码运行结果如下:

```
Student(Nancy,19,140116)
```

我们知道,如果不重写父类 toString 方法的话,返回的结果是类名加 hashcode 值,例如:

```
scala> class Person(var name:String,var age:Int)
defined class Person

scala> new Person("John",18)
res0: Person = Person@a21ce7
```

通过对父类方法的重写可以改变子类中的代码行为,例如:

```
scala> class Person(var name:String,var age:Int){
  //对父类 Any 中的 toString 方法进行重写
  override def toString = s"Person($name, $age)"
}
defined class Person

scala> new Person("John",18)
res1: Person = Person(John, 18)
```

6.4.4 多态

多态(Polymorphic)也称动态绑定(Dynamic Binding)或延迟绑定(Late Binding),指在执行期间而非编译期间确定所引用对象的实际类型,根据其实际类型调用其相应的方法,也就是说子类的引用可以赋给父类,程序在运行时根据实际类型调用对应的方法。请看下面的示例:

```
/**
 * 类继承:动态绑定与多态
 */
object Example6_11 extends App {

  //Person 类
  class Person(var name: String, var age: Int) {
    def walk(): Unit=println("walk() method in Person")
    //talkTo 方法,参数为 Person 类型
    def talkTo(p: Person): Unit=println("talkTo() method in Person")
```

```scala
}

class Student(name: String, age: Int) extends Person(name, age) {
  private var studentNo: Int = 0

  //重写父类的walk方法
  override def walk() = println("walk like an elegant swan")

  //重写父类的talkTo方法
  override def talkTo(p: Person) = {
    println("talkTo() method in Student")
    println(this.name + " is talking to " + p.name)
  }
}

class Teacher(name: String, age: Int) extends Person(name, age) {
  private var teacherNo: Int = 0

  //重写父类的walk方法
  override def walk() = println("walk like an elegant swan")

  //重写父类的talkTo方法
  override def talkTo(p: Person) = {
    println("talkTo() method in Teacher")
    println(this.name + " is talking to " + p.name)
  }
}

//下面的两行代码演示了多态的使用，Person类的引用可以指向Person类的任何子类
val p1: Person = new Teacher("albert", 38)
val p2: Person = new Student("john", 38)

//p1实际上引用的是Teacher类型的对象，Teacher类对父类中的walk方法进行了重写，因此
它调用的是重写后的方法
p1.walk()

//talkTo方法参数类型为Person类型，调用的是Student类重写后的talkTo方法
p1.talkTo(p2)
println("///////////////////////////////")

//p2引用的是Student类型的对象，Student类未对父类中的walk方法进行重写
//因此它调用的是继承自父类的walk方法
p2.walk()
```

```
//p2.talkTo(p1)传入的实际类型是Teacher,调用的是Teacher类重写后的talkTo方法
p2.talkTo(p1)
}
```

示例运行结果如下：

```
walk() method in Teacher
talkTo() method in Teacher
albert is talking to john
/////////////////////////
walk() method in Person
talkTo() method in Student
john is talking to albert
```

在Example6_11中定义了3个类，分别是Person、Student及Teacher类，Person类中定义了两个成员方法walk和talkTo，Student类和Teacher类继承自Person类，其中Student类对Person类中的talkTo方法进行了重写，而Teacher类对Person类中的walk和talkTo方法都进行了重写。代码val p1: Person = new Teacher("albert", 38)及val p2: Person = new Student("john", 38)创建了两个不同类型的对象，但p1和p2都为Person类型的引用，在运行p1.walk()及p2.walk()时，会根据引用的实际类型对象调用对应的方法，由于p1引用的是Teacher类型的对象，因此它调用的是Teacher类中的walk方法，而p2引用的是Student类型的对象，因此它调用的是Student类中的walk方法。

由于Teacher类对父类Person中的walk方法进行了重写，因此它的执行结果为"walk() method in Teacher"，而Student类没有对父类Person中的walk方法进行重写，因此它调用的是继承自父类的walk方法，从而其执行结果为"walk() method in Person"。子类Student和Teacher对父类中的talkTo方法都进行了重写，在调用p1.talkTo(p2)时，由于p1引用的是Teacher类型对象，p2引用的是Student类型的对象，因此在运行时调用的是Teacher类中的talkTo方法，而传入talkTo方法的具体对象类型为Student，因而其运行结果为"talkTo() method in Teacher, albert is talking to john"。p2.talkTo(p1)的程序运行逻辑与p1.talkTo(p2)类似。

6.5 成员访问控制

成员访问控制用于对类的成员访问权限进行控制，为说明Scala语言的成员访问控制方式，先介绍Java语言成员访问控制方式。Java语言中类成员的访问控制权限有4种，分别是package（默认）、private、public和protected。package为成员变量不加任何修饰时的访问权限，又叫做包访问，包访问允许成员变量和成员方法被同一个包内任何类的任何方法访问；private为私有访问，表示private修饰的成员变量和成员方法只能被同一个类中的其他方法访问；public为公有访问，表示public修饰的成员变量和成员方法在任何地方都可以访问；protected为保护访问，表示protected修饰的成员变量和成员方法在包内可以访问，包外部需要访问的话，需要该类为带有protected成员类的子类。Java四种不同访问控制方式的作用域如表6-1所示。

表 6-1 Java 四种不同成员访问控制的作用域

作用域	当前类	同一 package	子类	其他 package
public	√	√	√	√
protected	√	√	√	×
package	√	√	×	×
private	√	×	×	×

Scala 语言并没有直接使用 Java 语言的成员访问控制，而是自己实现了一套成员访问控制机制，其在成员访问控制方面比 Java 语言更灵活。Scala 的成员访问控制有 3 种，分别是默认的访问控制、private 访问控制、protected 访问控制。当不加任何关键字修饰成员时，使用的便是默认的访问控制，它对应 Java 语言中 public 关键字修饰的成员；private 和 protected 关键字修饰的成员与 Java 语言中的 private 和 protected 关键字修饰的成员作用域类似，只不过 private 和 protected 还可以同包一起使用来限制包中的访问控制权限，关于这部分内容将放在 8.2 节中介绍。 本节将对默认的访问控制、protected 访问控制、private 及 private[this]这四种成员访问控制方式进行介绍。

6.5.1 默认访问控制

Scala 中如果在成员变量或方法前面不加任何访问控制符，则其等同于 Java 中的 public 关键字，Scala 中没有 public 这个关键字，请看下面的例子。

```
scala> class Person {
  var name:String=null
  def print=println(s"$name")
}
defined class Person

scala> val p=new Person
p: Person = Person@1bd9675

//直接访问成员变量 name
scala> p.name="John"
p.name: String = John

//直接访问成员方法 print
scala> p.print
John
```

可以看到，默认访问控制与 Java 语言中的 public 关键字作用相同，成员变量与成员方法在类的内部和外部都可以直接进行访问。这种情况对在主构造函数中定义的成员变量同样适用，例如：

```
/主构造函数中的定义成员变量
scala> class Person(var name:String)
```

```
defined class Person

scala> val p=new Person("John")
p: Person = Person@11fe5db

scala> p.name
res0: String = John

scala> p.name="john"
p.name: String = john

scala> p.name
res1: String = john
```

现在我们对默认访问控制所生成的 getter 与 setter 方法及作用域进行总结，如表 6-2 所示。

表 6-2 默认访问控制成员域生成的方法及访问范围

成员域	生成的方法	作用域
val name	只生成 public name 方法	在类、子类及外部都可以访问
var name	生成 public name 和 public name_=方法	在类、子类及外部都可以访问
@BeanProperty val name	生成 public name 及 public getName 方法	在类、子类及外部都可以访问
@BeanProperty var name	生成 public name、 public name_=、public getName 及 public setName 方法	在类、子类及外部都可以访问

6.5.2 protected 访问控制

Protected 访问控制符修饰的类成员只能在该类及其子类中访问，在外部不能访问，具体代码如下：

```
scala> class Person {
 //protected 成员变量 name
  protected  var name:String=null
  def print=println(s"$name")
}
defined class Person

scala> class Student extends Person{
  //子类中能够访问成员变量 name
  override def toString: String = this.name
}
defined class Student


scala> val p=new Person
p: Person = Person@b71be
```

```
//外部不能访问
scala> p.name
<console>:10: error: variable name in class Person cannot be accessed in Person
  Access to protected method name not permitted because
  enclosing object $iw is not a subclass of
  class Person where target is defined
          p.name
```

从上面的代码不难看出，Person 类中的成员变量 name 使用了 protected，因此成员变量 name 只能在 Person 类及其子类中使用，在外部不能访问。对 Person 类生成的字节码文件进行反编译后可以得到下述代码：

```
E:\IntellijIDEAWorkspace\out\production\ScalaProject\cn\scala\chapter06>javap -private Person.class
Compiled from "Person.scala"
public class cn.scala.chapter06.Person {
  private java.lang.String name;
  public java.lang.String name();
  public void name_$eq(java.lang.String);
  public void print();
  public cn.scala.chapter06.Person();
}
```

可以看到，虽然成员变量 name 被声明为 protected，但它生成的 Scala 风格的 getter、setter 方法与默认访问控制一样，不同的是作用域范围更小，见表 6-3 所示。

表 6-3 protected 成员域生成的方法及访问范围

成员域	生成的方法	作用域
protected val name	public name 方法	在类、子类可以访问
protected var name	public name、public name_= 方法	在类、子类可以访问
@BeanProperty protected val name	public name、public getName 方法	在类、子类可以访问
@BeanProperty protected var name	public name、public name_=、public getName 及 public setName 方法	在类、子类可以访问

6.5.3 private 访问控制

由 private 关键字修饰的类成员只能在内部被该类的成员访问，在外部及其子类中不能被访问。例如：

```
scala> class Person {
  //private 关键字修饰的成员变量
  private var name:String=null
```

```
  //只能被Person类的成员访问
  def print=println(s"$name")
}
defined class Person

//在子类中不能访问父类的private成员
scala> class Student extends Person{
  override def toString: String = this.name
}
<console>:9: error: variable name in class Person cannot be accessed in Student
        override def toString: String = this.name

//外部不能访问
scala> p.name
<console>:10: error: variable name in class Person cannot be accessed in Person
        p.name
```

对 Person 类生成的字节码文件进行反编译可以得到下述代码：

```
E:\IntellijIDEAWorkspace\out\production\ScalaProject\cn\scala\chapter06>javap -private Person.class
Compiled from "Person.scala"
public class cn.scala.chapter06.Person {
  private java.lang.String name;
  private java.lang.String name();
  private void name_$eq(java.lang.String);
  public void print();
  public cn.scala.chapter06.Person();
}
```

可以看到，private 修饰的成员变量 name，最终生成的 Scala 风格 getter 方法 name()、setter 方法 name_$eq(java.lang.String)也为 private。需要注意的是，@BeanProperty 只能作用于非 private 成员变量，例如：

```
scala> import scala.beans.BeanProperty
class Person {
  //@BeanProperty只能作用于非private成员变量
  @BeanProperty private var name:String=null
  def print=println(s"$name")
}
<console>:9: error: `BeanProperty' annotation can be applied only to non-private fields
        @BeanProperty private var name:String=null
```

通过上面的代码可以看到，@BeanProperty 不能作用于 private 成员变量。private 成员访问控制权限如表 6-4 所示。

表 6-4　private 成员域生成的方法及访问范围

成员域	生成的方法	作用域
private val name	private name 方法（定义了伴生对象，则为 public name）	只能在本类及伴生对象中访问
private var name	private name（定义了伴生对象，则为 public name）、private name_=方法	只能在本类及伴生对象中访问
@BeanProperty private val name	——	——
@BeanProperty private var name	——	——

6.5.4　private[this]访问控制

在 Scala 中还存在 private[this]关键字修饰的类成员，这种类成员被称为类的私有成员，先看下列示例代码。

```
scala> class Person{
  //private[this]修饰的成员变量为类的私有成员
  private[this] var name:String=_
  def this(name:String)={
    this();
    this.name=name
  }
  //类的私有成员只能在类内部访问
  def print=println(name)
}
defined class Person

scala> val p=new Person("John")
p: Person = Person@10ef83c

//类的私有成员，外部不能访问
scala> p.name
<console>:10: error: value name is not a member of Person
          p.name

scala> p.print
John
```

在 Person 类中，通过 private[this]对成员变量 name 进行修饰，表示该成员变量为类的私有成员，只能在类内部使用，本例中 def print=println(name)访问的便是该类的私有成员，而在类外部不能访问，例子中通过 p.name 访问成员变量 name 会出错。但细心的读者可能会发现，priate 和 private[this]之间似乎并没有什么特别的不同，其实它们两者之间的最主要的差别在于伴生对象的访问权限。

```scala
/**
 * private 关键字修饰的成员变量:伴生类与伴生对象中可以访问
 * */
object Example6_15 extends App {
  class Person{
    //private 修饰的成员变量
    private var name:String=_
    def this(name:String){
      this();
      this.name=name
    }
    //在本类中可以访问
    def print=println(name)
  }

  object Person{
    //在伴生对象中可以访问伴生类对象的 private 成员变量
    def printName=println(new Person("John").name)
  }

  Person.printName
  //在伴生对象和伴生类的外部不能访问 private 成员变量
  //println(new Person("John").name)
}
```

通过上述代码可以看到，private 关键字修饰的成员变量在伴生类和伴生对象中可以被访问，在外部不能访问。但如果伴生类中的成员变量被 private[this]关键字修饰，则只能在伴生类中被访问，在伴生对象及外部都不能被访问。

```scala
/**
 * private[this]修饰的成员变量，只能在伴生类中访问，在伴生对象及外部不能访问
 * */
object Example6_16 extends App {
  class Person{
    //private[this]修饰的成员变量
    private[this] var name:String=_
    def this(name:String){
      this();
      this.name=name
    }
```

```
    //在本类中可以访问
    def print=println(name)
  }

  object Person{
    //在伴生对象中不能访问
    //def printName=println(new Person("John").name)
  }

  //在外部不能访问private[this]成员变量
  //println(new Person("John").name)
}
```

在 Person 类当中，成员变量 name 使用 private[this]修饰，只能在伴生类中访问（见代码 def print=println(name)），在伴生对象中不能访问（见代码 def printName=println(new Person("John").name)，在外部也不能访问（见代码 println(new Person("John").name)），这就是为什么 private[this]修饰的成员变量被称为对象私有成员变量的原因，因为它只能在对象的内部使用。

为了进一步说明 private 与 private[this]两者的区别，我们来对不同情况下类生成的字节码文件进行反编译，来对比相应的类生成的成员变量和方法的访问控制权限。无伴生对象、类成员变量为 private 时的代码如下：

```
/**
  * 无伴生对象，成员变量为private
  **/
object Example6_17 extends App {
  class Person{
    //private 修饰的成员变量
    private var name:String=_
    def this(name:String){
      this();
      this.name=name
    }
  }
}
```

Person 类生成的字节码反编译后的结果如下：

```
E:\IntellijIDEAWorkspace\out\production\ScalaProject\cn\scala\chapter06>javap -private Example6_17$Person.class
Compiled from "Example6_01.scala"
public class cn.scala.chapter06.Example6_17$Person {
  private java.lang.String name;
  private java.lang.String name();
  private void name_$eq(java.lang.String);
```

```
    public cn.scala.chapter06.Example6_17$Person();
    public cn.scala.chapter06.Example6_17$Person(java.lang.String);
}
```

有伴生对象，类成员变量为 private 的代码如下：

```
/**
  * 有伴生对象，成员变量为private
  * */
object Example6_18 extends App {
  class Person{
    //private 修饰的成员变量
    private var name:String=_
    def this(name:String){
      this();
      this.name=name
    }
  }
  object Person{
    //在伴生对象中可以访问伴生类对象的private成员变量
    def printName=println(new Person("John").name)
  }
}
```

Person 类的字节码文件反编译后的结果如下：

```
E:\IntellijIDEAWorkspace\out\production\ScalaProject\cn\scala\chapter06>ja
vap -private Example6_18$Person.class
Compiled from "Example6_01.scala"
public class cn.scala.chapter06.Example6_18$Person {
  private java.lang.String cn$scala$chapter06$Example6_18$Person$$name;
  public java.lang.String cn$scala$chapter06$Example6_18$Person$$name();
  private void cn$scala$chapter06$Example6_18$Person$$name_$eq(java.lang.S
tring)
;
  public cn.scala.chapter06.Example6_18$Person();
  public cn.scala.chapter06.Example6_18$Person(java.lang.String);
}
```

无伴生对象，有 private[this] 成员变量的代码如下：

```
/**
  * 无伴生对象，有private[this]成员变量
  * */
object Example6_19 extends App {
  class Person{
```

```scala
    //private[this]修饰的成员变量
    private[this] var name:String=_
    def this(name:String){
      this();
      this.name=name
    }
  }
}
```

Person 类字节码文件反编译后得到的代码如下：

```
E:\IntellijIDEAWorkspace\out\production\ScalaProject\cn\scala\chapter06>ja
vap -private Example6_19$Person.class
Compiled from "Example6_01.scala"
public class cn.scala.chapter06.Example6_19$Person {
  private java.lang.String name;
  public cn.scala.chapter06.Example6_19$Person();
  public cn.scala.chapter06.Example6_19$Person(java.lang.String);
}
```

有伴生对象，有 private[this]成员变量的代码如下：

```scala
/**
  * 有伴生对象，有private[this]成员变量
  **/
object Example6_20 extends App {

  class Person{
    //private[this]修饰的成员变量
    private[this] var name:String=_
    def this(name:String){
      this();
      this.name=name
    }
  }
  object Person{

  }
}
```

Person 类生成的字节码文件反编译后的代码如下：

```
E:\IntellijIDEAWorkspace\out\production\ScalaProject\cn\scala\chapter06>ja
vap -private Example6_20$Person.class
Compiled from "Example6_01.scala"
public class cn.scala.chapter06.Example6_20$Person {
```

```
    private java.lang.String name;
    public cn.scala.chapter06.Example6_20$Person();
    public cn.scala.chapter06.Example6_20$Person(java.lang.String);
}
```

通过上面的代码可以得到 private[this]成员域生成的方法及访问范围，如表 6-5 所示。

表 6-5 private[this]成员域生成的方法及访问范围

成员域	生成的方法	作用域
private[this] val name	——	只能在本类中访问，在伴生对象及外部不能访问
private[this] var name	——	只能在本类中访问，在伴生对象及外部不能访问
@BeanProperty private val name	——	——
@BeanProperty private var name	——	——

6.5.5 主构造函数中的成员访问控制

主构造函数中定义成员变量时，也可以加成员访问控制符，在讲解主构造函数时我们曾经使用过下列代码

```
class Person(var name:String,var age:Int)
```

定义类、成员变量及主构造函数，这种方式会自动帮我们生成 private 成员变量、public getter 和 setter 方法，也就是在定义主构造函数时不加任何的修饰符，默认其访问控制权限与 Java 语言的 public 关键字对应。

接下来对主构造函数中被 protected、private 及 private[this]修饰的成员变量的访问权限进行介绍。

（1）protected 访问控制符

```
scala> //主构造函中定义的成员变量 name 使用 protected 关键字修饰，而 age 使用默认访问控制符
    class Person(protected var name:String,var age:Int)
    class Student(name:String,age:Int,var studentNo:String) extends Person(name,age)
    defined class Person
    defined class Student

scala> val p=new Person("John",18)
p: Person = Person@1653cd6

//不能外部直接访问 proteted 成员
```

```
scala> p.name="john"
<console>:12: error: variable name in class Person cannot be accessed in Person
 Access to protected method name not permitted because
 enclosing object $iw is not a subclass of
 class Person where target is defined
val $ires0 = p.name
             ^
//默认控制符的成员可以直接访问
scala> p.age=19
p.age: Int = 19
```

通过上述代码可以看到，通过 class Person(protected var name:String,var age:Int)这种带主构造函数的类定义方式，其访问控制权限等同于下列代码：

```
class Person{
   protected var name:String=_
   var age:Int=_
   def this(name:String,age:Int){
     this()
     this.name=name
     this.age=age
   }
 }
```

（2）private 访问控制符

主构造函数中定义 private 成员变量的示例代码如下：

```
//主构造函数中的参数类型为private
scala> class Person(private var name:String, private var age:Int)
defined class Person

scala> val p=new Person("John",18)
p: Person = Person@7d9cbb

//外部不能访问
scala> p.name
<console>:10: error: variable name in class Person cannot be accessed in Person
          p.name
          ^
```

对该 Person 类生成的字节码文件进行反编译可以看到如下代码。

```
E:\IntellijIDEAWorkspace\out\production\ScalaProject\cn\scala\chapter06>ja
```

```
vap -private Person.class
    Compiled from "Example6_01.scala"
    public class Person {
      private java.lang.String name;
      private int age;
      private java.lang.String name();
      private void name_$eq(java.lang.String);
      private int age();
      private void age_$eq(int);
      public Person(java.lang.String, int);
    }
```

在上述代码中，可以看到主构造函数中的 private 类型参数最终会产生对应的 private 类成员变量和 private getter 与 setter 方法。主构造函数中 private 类型参数最终生成的类成员变量、方法及访问权限与在类中定义的 private 类成员变量访问权限是一致的，具体可以参考表 6-3。

（3）private[this] 访问控制符

在类中定义的成员变量使用 private[this] 修饰，该成员变量是对象私有成员变量，只能在类的内部使用，在伴生对象及外部不能使用。在主构造函数中也同样适用，只不过在主构造函数中定义私有成员变量，可以不用 private[this] 关键字。

```
//主构造函数参数没有使用 val 或 var 关键字定义时，默认为 private[this]
scala> class Person(name:String, age:Int){
  override def toString=s"Student($name,$age)"
}
defined class Person

scala> val p=new Person("John",18)
p: Person = Student(John,18)

//外部不能访问
scala> p.name
<console>:10: error: value name is not a member of Person
          p.name
          ^
```

对 Persond 类生成的字节码文件进行反编译后得到如下代码。

```
E:\IntellijIDEAWorkspace\out\production\ScalaProject\cn\scala\chapter06>ja
vap -private Person.class
    Compiled from "Example6_01.scala"
    public class cn.scala.chapter06.Person {
      private final java.lang.String name;
      private final int age;
      public java.lang.String toString();
```

```
        public cn.scala.chapter06.Person(java.lang.String, int);
}
```

在上述代码中可以看到，最后生成的类成员变量都为 private final 类型，而且没有生成相应的 getter 与 setter 方法。也就是说代码 class Person(name:String, age:Int)等同于代码 class Person(private[this] val name:String,private[this] val age:Int)，示例如下。

```
//与代码class Person(name:String, age:Int)等同
scala> class Person(private[this] val name:String,private[this] val age:Int){
  override def toString=s"Student($name,$age)"
}
defined class Person

scala> val p=new Person("John",18)
p: Person = Student(John,18)

scala> p.name
<console>:10: error: value name is not a member of Person
         p.name
         ^
```

感兴趣的读者可以将

```
class Person(private[this] val name:String,private[this] val age:Int){
  override def toString=s"Student($name,$age)"
}
```

类生成的字节码文件进行反编译去验证。如果需要定义非 val 类型的对象私有成员变量，只能通过下列代码进行操作：

```
class Person(private[this] var name:String, private[this] var age:Int){
    override def toString=s"Student($name,$age)"
}
```

在主构造函数中定义对象私有成员变量时，如果类中没有类成员使用该成员变量，则不会生成该私有成员变量，例如：

```
//类中没有类成员使用主构造函数中定义的对象私有成员变量
class Person(name:String, age:Int)
```

相比于前面的 Person 类，类中已经没有 toString 方法，因此类中没有任何类成员如成员方法、内部类使用对象私有变量 name 和 age，则这两个成员变量不会生成。对 Person 类生成的字节码文件进行反编译后得到如下代码。

```
E:\IntellijIDEAWorkspace\out\production\ScalaProject\cn\scala\chapter06>javap -private Example6_21$Person.class
Compiled from "Example6_01.scala"
```

```
public class cn.scala.chapter06.Example6_21$Person {
  public cn.scala.chapter06.Example6_21$Person(java.lang.String, int);
}
```

通过上述代码可以看到，反编译后的代码中并没有 private 类型的成员变量。

6.6 抽象类

6.6.1 抽象类的定义

抽象类是一种不能被实例化的类，抽象类中存在抽象成员变量或成员方法，这些成员方法或成员变量在子类中被具体化。在 Java 语言中，抽象类使用关键字 abstract 来声明，Scala 语言同样通过 abstract 关键字定义抽象类，Scala 语言中的抽象类不仅可以有抽象方法，还可以有抽象成员变量。Scala 中一般类的成员变量在定义时必须初始化。例如：

```
//普通类成员变量在定义时必须显式初始化
scala> class Person{
  var name:String=_
}
defined class Person

//不显式初始化成员变量会报错，提示 Person 类应被声明为 abstract
scala> class Person{
  var name:String
}
<console>:7: error: class Person needs to be abstract, since variable name is not defined
(Note that variables need to be initialized to be defined)
       class Person{
             ^
```

可以看到，当普通类的成员变量不显式地对其进行初始化时会报错并提示类应该被声明为 abstract，这说明 Scala 语言中的抽象类可以有抽象成员变量（也叫抽象字段）。

```
//成员变量如果不显式初始化，则将类声明为抽象类，通过 abstract 关键字来定义
scala> abstract class Person{
  var name:String
}
defined class Person
```

当子类继承抽象类时，需要在子类中对父类中的抽象成员变量进行初始化，否则子类也必须声明为抽象类。

//父类中包含抽象成员变量时，子类如果为普通类则必须将该成员变量初始化，否则子类也应声明为

抽象类

```
scala> class Student extends Person
<console>:8: error: class Student needs to be abstract, since variable name in class Person of type String is not defined
(Note that variables need to be initialized to be defined)
       class Student extends Person
             ^
//在子类中对父类中的抽象成员变量进行初始化,使用 override 关键字
scala> class Student extends Person{
  override var name:String=_
}
defined class Student

//也可以省略 override 关键字
scala> class Student extends Person{
  var name:String=_
}
defined class Student
```

通过上述代码可以看到,如果 Person 类中存在抽象成员变量 name,子类 Student 如果没有对该成员变量进行初始化的话,系统会报错并提示应该将子类也初始化 abstract,如果不需要子类为抽象类,则需要对该成员变量进行初始化,值得注意的是子类对父类抽象成员变量进行重写可以加 override 关键字也可省略。

6.6.2 抽象类的使用

前面给出的是包含抽象成员变量的 Scala 抽象类,Scala 中的抽象类还可以定义相应的抽象方法、具体方法及具体成员变量,例如:

```
/**
 * 抽象类
 */
object Example6_12 extends App {
  abstract class Person{
    //抽象类中的具体成员变量
    var age:Int=0
    //抽象类中的抽象成员变量
    var name:String
    //抽象类中的抽象方法
    def walk()
    //抽象类中的具体方法
    override def toString=name
  }
```

```scala
class Student extends Person{
  //对继承自父类的抽象成员变量进行初始化,override 关键字可以省略
  override var name: String = _

  //对继承父类的抽象成员方法进行初始化,override 关键字可以省略
  override def walk(): Unit = println("Walk like a Student")
}

val p=new Student
p.walk()
```

通过上面的代码可以看到,Person 类中不但包含了抽象成员变量,还包含了具体成员变量、抽象成员方法和具体成员方法,子类 Student 在继承 Person 类时,必须对父类中的抽象成员变量进行初始化及对抽象成员方法进行实现,否则子类也必须声明为抽象类,在对抽象成员变量和抽象成员方法进行 override 时,关键字 override 可以省略。对 Person 类及 Student 类生成的字节码文件进行反编译得到如下代码。

```
E:\IntellijIDEAWorkspace\out\production\ScalaProject\cn\scala\chapter06>javap -private Example6_12$Person.class
Compiled from "Example6_01.scala"
public abstract class cn.scala.chapter06.Example6_12$Person {
  private int age;
  public int age();
  public void age_$eq(int);
  public abstract java.lang.String name();
  public abstract void name_$eq(java.lang.String);
  public abstract void walk();
  public java.lang.String toString();
  public cn.scala.chapter06.Example6_12$Person();
}

E:\IntellijIDEAWorkspace\out\production\ScalaProject\cn\scala\chapter06>javap -private Example6_12$Student.class
Compiled from "Example6_01.scala"
public class cn.scala.chapter06.Example6_12$Student extends cn.scala.chapter06.Example6_12$Person {
  private java.lang.String name;
  public java.lang.String name();
  public void name_$eq(java.lang.String);
  public void walk();
  public cn.scala.chapter06.Example6_12$Student();
}
```

通过上面反编译后的代码可以看到，Scala 抽象类最终会对应生成一个 Java 抽象类，Scala 中的抽象成员变量 name 在生成的 Java 抽象类中只有对应的 getter、setter 抽象方法而不会有成员变量 name，Scala 中的具体成员变量 age 在生成的 Java 抽象类中不但有对应的 getter、setter 具体成员方法还会生成相应的成员变量 age。抽象成员方法在生成的 Java 抽象类中是抽象的，具体成员方法 toString 在生成的 Java 抽象类中是具体的。子类 Student 对父类中的抽象成员变量和抽象成员方法进行 override 后，在子类中会生成相应的成员变量及对应的方法。

6.7 内部类与内部对象

内部类指的是定义在对象或类内部的类，内部对象则指的是定义在对象或类内部的对象。在本章的很多示例中都大量使用了内部类，如我们在应用程序对象 Example6_12 中定义了内部类 Person 和 Student。下列示例代码 Example6_13 给出的是内部类和内部对象的使用，Example6_13 本身是一个单例对象，class Student 和 object Student 为定义在应用程序对象 Example6_13 中的内部类和内部对象，然后在 class Student 和 object Student 内部又分别定义了相应的内部类和内部对象。

```
/**
 * 内部类与对象
 */
object Example6_13 extends App {

  //伴生类 Student
  class Student(var name:String,var age:Int){
    //内部类 Grade
    class Grade(var name:String)

    //内部对象
    object Utils1{
      def print(name:String)=println(name)
    }
  }

  //伴生对象 Student
  object Student{
    //单例对象中的内部类 Printer
    class Printer{
      def print(name:String)=println(name)
    }

    //伴生对象中的单例对象
```

```scala
    object Utils2{
      def print(name:String)=println(name)
    }
  }

  val student=new Student("John",18)
  //创建伴生类的内部类对象
  val grade=new student.Grade("大学一年级")
  println("new student.Grade(\"大学一年级\").name：调用伴生类的内部类方法, grade.name="+grade.name)

  //调用伴生类的内部对象方法
  student.Utils1.print("student.Utils1.print：调用伴生类的内部对象方法")

  //创建伴生对象的内部类对象
  val printer=new Student.Printer()
  printer.print("new Student.Printer().print：调用伴生对象的内部类方法")

  //调用伴生对象的内部对象方法
  Student.Utils2.print("Student.Utils2.print：调用伴生对象的内部对象方法")

}
```

代码运行结果如下：

```
new student.Grade("大学一年级").name：调用伴生类的内部类方法, grade.name=大学一年级
student.Utils1.print：调用伴生类的内部对象方法
new Student.Printer().print：调用伴生对象的内部类方法
Student.Utils2.print：调用伴生对象的内部对象方法
```

在 Example6_13 中，定义了伴生类 Student 和伴生对象 Student，在伴生类 Student 中定义了一个内部类 Grade 和内部对象 Utils1，在伴生对象 Student 中则定义了内部类 Printer 和内部对象 Utils。创建内部类 Grade 的对象时，使用代码 val grade=new student.Grade("大学一年级")，即通过外部类对象.内部类名称（内部类构造函数参数）的方式创建内部类对象，可以看到访问内部类的方式如同内部类是其成员变量一样。通过代码 student.Utils1.print，访问伴生类 Student 的内部对象并调用其方法，即通过外部类对象.内部对象的方式访问内部对象。使用代码 val printer=new Student.Printer()创建伴生对象的内部类对象，即通过伴生对象.内部类名称（内部类构造函数参数）的方式创建伴生对象的内部类对象。对于伴生对象的内部对象访问，则通过伴生对象.内部对象的方式来访问，代码 Student.Utils2.print 便是如此。

6.8 匿名类

匿名类，顾名思义就是没有名字的类。当某个类在代码中仅使用一次时，可以考虑使用匿名类，示例如下：

```scala
/**
  * 匿名类
  */
object Example6_14 extends App {

  abstract class Person(var name:String,var age:Int){
    def print:Unit
  }

  //使用匿名类,并创建匿名类对象
  val p=new Person("John",18) {
    override def print: Unit = println(s"Person($name,$age)")
  }

  p.print

}
```

代码运行结果如下：

```
Person(John,18)
```

代码

```scala
val p=new Person("John",18) {
    override def print: Unit = println(s"Person($name,$age)")
}
```

使用的便是匿名类创建对象，上面的这几行代码等同于下面的代码：

```scala
class NamedClass(name:String,age:Int) extends Person(name,age){
    override def print: println(s"Person($name,$age)")
}
val p=new NamedClass("John",18)
```

只不过，命名类 NamedClass 一旦被定义就可以反复使用，而匿名类只使用一次，代码更简洁。事实上匿名类会生成对应的字节码文件 **Example6_14$$anon$1.class**，对其进行反编译可以看到如下代码：

```
E:\IntellijIDEAWorkspace\out\production\ScalaProject\cn\scala\chapter06>javap -private Example6_14$$anon$1.class
Compiled from "chapter06.scala"
```

```
public final class cn.scala.chapter06.Example6_14$$anon$1 extends cn.scala.
chapter06.Example6_14$Person {
    public void print();
    public cn.scala.chapter06.Example6_14$$anon$1();
}
```

也就是说，匿名类仍然会单独生成相应的字节码文件。

小　结

本章介绍了 Scala 的面向对象编程特性，内容涉及 Scala 类的定义、对象的创建，主构造函数的定义、作用与使用方法，辅助构造函数的定义、作用与使用方法，Scala 类的继承，成员变量及成员方法访问控制，抽象类的定义与使用、内部类与对象、匿名类等。在关键特性的理解上，通过与 Java 语言中面向对象编程相关内容相比较，来加深对 Scala 面向对象编程的理解。在下一章中，将会对 Scala 面向对象编程中另外一个重要的特性——trait 进行介绍。

Scala 面向对象编程（下）

trait 是 Scala 面向对象编程中的一个重要内容，它与 Java 语言中的接口、抽象类有很多的相似之处，但 trait 提供了更强大、灵活的使用方式。本章将介绍 trait 的定义与使用、trait 与类的关系、多重继承等重要内容。

7.1　trait 简介

在 Scala 语言中并没有提供 Java 语言的 interface 关键字来定义接口，而是提供 trait（中文常译为特质）关键字来封装成员方法和成员变量，在使用时通过 extends 或 with 关键字来混入（mix）定义的 trait。trait 与类继承最大的不同在于一个类只能有一个父类但可以混入多个 trait。

先看下面的例子：

```
//使用trait关键字定义一个名为Closable的特质
scala> trait Closable{
  def close():Unit
}
defined trait Closable
```

代码中的 trait 关键字定义了一个名为 Closable 的 trait，并在该 trait 中定义了一个抽象方法。假设现在有一个文件类 File，它需要实现自己的 close 方法以便关闭文件，此时可以通过关键字 extends 将 Closable trait 混入，并对 close 方法进行实现。

```
//使用extends关键字将Closable trait混入，并实现close方法
scala> class File(var name:String) extends Closable{
  def close():Unit=println(s"File $name has been closed")
}
defined class File

scala> new File("config.txt").close()
File config.txt has been closed
```

代码 class File(var name:String) extends Closable 定义一个 File 类并混入了 Closable trait，然后在 File 类中对 close 方法进行实现。熟悉类继承的读者可能会发现，这里的 trait 似乎跟抽象类并没有太多的区别，其实完全可以定义一个抽象类 Closable，并在类中定义一个抽象方法 close，然后再通过 File 类继承该抽象类。事实上，它们背后的机理是完全不同的，对 Closable 生成的字节码文件进行反编译可以看到如下代码：

```
E:\IntellijIDEAWorkspace\out\production\ScalaProject\cn\scala\chapter07>javap -private Example07_01$Closable.class
Compiled from "chapter07.scala"
public interface cn.scala.chapter07.Example07_01$Closable {
  public abstract void close();
}
```

而对混入 Closable 的 File 类生成的字节码文件进行反编译后可以看到如下代码：

```
E:\IntellijIDEAWorkspace\out\production\ScalaProject\cn\scala\chapter07>javap -private Example07_01$File.class
Compiled from "chapter07.scala"
```

```
public class cn.scala.chapter07.Example07_01$File implements cn.scala.chap
ter07.Example07_01$Closable {
  private java.lang.String name;
  public java.lang.String name();
  public void name_$eq(java.lang.String);
  public void close();
  public cn.scala.chapter07.Example07_01$File(java.lang.String);
}
```

从反编译后的结果可以看到，代码中定义的 Closable 特质最终生成的字节码对应的是 Java 语言中的 interface，而 File 类通过 implements 关键字对该 interface 进行实现。

当 File 类既继承其他类，又需要混入若干特质时，则需要使用 with 关键字，例如：

```
//File 继承自 java.io.File 类，并混入 Closable 和 Cloneable 两个特质
scala> class File(var name:String) extends java.io.File(name) with Closable
with Cloneable{
  def close()=println(s"File $name has been closed")
}
defined class File
```

代码 class File(var name:String) extends java.io.File(name) with Closable with Cloneable 定义的 File 类使用 extends 关键字扩展 java.io.File 类，然后通过 with 关键字混入前面定义的 Closable trait 和 Scala 自身提供的 Cloneable trait。Cloneable 扩展自 java.lang.Cloneable 接口，它被定义在 scala 包中，例如：

```
package scala

/**
 * Classes extending this trait are cloneable across platforms (Java, .NET).
 */
trait Cloneable extends java.lang.Cloneable
```

这里使用的是 extends 关键字定义 Cloneable。可以看到，trait 的定义有两种方式：

（1）直接通过 trait 关键字进行定义，如前面定义的 Closable。

（2）同类的定义一样，使用 extends 关键字对现有的 Java 接口进行扩展（如 scala.Cloneable），或对现有的 scala trait 进行扩展，混入的第一个 trait 使用 extends 关键字，其他的 trait 则使用 with 关键字。

```
// Processable 通过 extends 关键字和 with 关键字混入 Closable、Cloneable 两个 trait
scala> trait Processable extends Closable with Cloneable
defined trait Processable

//混入的第一个 trait 必须使用 extends 关键字
scala> trait Processable with Closable with Cloneable
<console>:1: error: ';' expected but 'with' found.
```

```
trait Processable with Closable with Cloneable
                    ^
```

一个类混入的 trait 如果有抽象方法，混入后应该对 trait 中的抽象方法进行实现，否则的话该类应该声明为抽象类。

```
//不实现 trait 中的抽象方法，则需要将定义的类定义为抽象类
scala> abstract class File(var name:String) extends java.io.File(name) with Closable with Cloneable
defined class File
```

7.2　trait 的使用

7.2.1　trait 的几种不同用法

trait 与抽象类类似，在 trait 中可以有抽象成员、成员方法、具体成员和具体方法，下面对这几种类型的 trait 的使用进行介绍。

（1）只有抽象方法的 trait

只有抽象方法的 trait 在使用上类似于 Java 语言中的 interface，下面的代码演示的是只有抽象方法的 trait 的使用方法。

```
/**
 * 只有抽象方法的 trait
 */
object Example07_01 extends App{
  case class Person(var id:Int,var name:String,var age:Int)

  //定义只包含抽象方法的 trait,
  trait PersonDAO{
    //添加方法
    def add(p:Person)
    //更新方法
    def update(p:Person)
    //删除方法
    def  delete(id:Int)
    //查找方法
    def findById(id:Int):Person
  }

  class PersonDAOImpl extends PersonDAO {
    //添加方法
    override def add(p: Person): Unit = {
```

```
      println("Invoking add Method....adding "+p)
    }
    //更新方法
    override def update(p: Person): Unit = {
      println("Invoking update Method,updating "+p)
    }
    //查找方法
    override def findById(id: Int): Person = {
      println("Invoking findById Method,id="+id)
      Person(1,"John",18)
    }
    //删除方法
    override def delete(id: Int): Unit = {
      println("Invoking delete Method,id="+id)
    }
  }

  val p:PersonDAO=new PersonDAOImpl
  p.add(Person(1,"John",18))
}
```

（2）只有抽象成员变量和成员方法的 trait

trait 中还可以定义抽象字段，例如：

```
//带抽象成员变量的 trait,
trait PersonDAO{
  //抽象成员
  var recordNum:Long
  //添加方法
  def add(p:Person)
  //更新方法
  def update(p:Person)
  //删除方法
  def  delete(id:Int)
  //查找方法
  def findById(id:Int):Person
}
```

对 PersonDAO 生成的字节码文件进行反编译得到下列代码：

```
public interface cn.scala.chapter07.Example07_02$PersonDAO {
  public abstract long recordNum();
  public abstract void recordNum_$eq(long);
  public abstract void add(cn.scala.chapter07.Example07_02$Person);
  public abstract void update(cn.scala.chapter07.Example07_02$Person);
  public abstract void delete(int);
```

```
public abstract cn.scala.chapter07.Example07_02$Person findById(int);
}
```

可以看到，与只有抽象成员方法的 trait 一样，它最终的实现与 Java 语言中的 interface 等价，抽象成员变量最终会生成抽象的 getter 方法和 setter 方法。

（3）有具体成员变量的 trait

trait 中还可以有具体成员变量，例如：

```
//带具体成员变量的trait
trait PersonDAO{
  //具体成员
  var recordNum:Long=_
  //添加方法
  def add(p:Person)
  //更新方法
  def update(p:Person)
  //删除方法
  def  delete(id:Int)
  //查找方法
  def findById(id:Int):Person
}
```

与只有抽象成员变量和抽象成员方法的 trait 所不同的是，PersonDAO 最终会生成两个字节码文件，分别是 PersonDAO$class.class 和 PersonDAO.class，感兴趣的读者可以查看它们字节码反编译后的代码。

（4）有具体成员方法的 trait

trait 还可以有具体成员方法，例如

```
//带具体成员的trait
trait PersonDAO{
  //具体成员变量
  var recordNum:Long=_
  //具体成员方法
  def add(p:Person):Unit = {
    println("Invoking add Method....adding "+p)
  }
  //更新方法
  def update(p:Person)
  //删除方法
  def  delete(id:Int)
  //查找方法
  def findById(id:Int):Person
}
```

与带具体成员变量的 trait 一样，这种语法特性 PersonDAO 会生成 PersonDAO$class.class 和 PersonDAO.class 两个字节码文件。

通过前面 trait 的四种不同用法，熟悉 Java 语言的读者们会发现：相比于 Java 语言中的接口和抽象类，Scala 中的 trait 与 Java 中的抽象类更接近，因为 Java 中的抽象类允许类中有抽象成员变量、具体成员变量、抽象成员方法及具体成员方法，而接口中允许有抽象成员变量和具体成员方法，例如：

```
//Java 语言中的抽象类
abstract class A{
    //抽象类中的具体成员变量
    private String a="String";
    //抽象类中的抽象成员变量
    private int b;
    //抽象类中的抽象成员方法
    public abstract void print();
    //抽象类中的具体成员方法
    public void println(){
        System.out.println("a");
    }
}

//Java 语言中的接口
interface B{
    String a="String";

    //不允许有抽象成员变量
    //int b;

    public abstract void print();

    //不允许有具体成员方法
    /*public void println(){
        System.out.println("a");
    }*/
}
```

至于 trait 与类之间的异同，将在 7.3 节中讨论。

7.2.2 混入 trait 的类对象构造

在定义类时，如果在类体中有执行语句，则在创建对象时会执行这些语句，在 trait 中同样可以加入执行语句，创建对象时便会自动执行这些语句，例如：

```
/**
 * trait 构造顺序
 */
object Example07_05 extends App{
```

```
import java.io.PrintWriter
trait Logger{
  println("Logger")
  def log(msg:String):Unit
}

trait FileLogger extends Logger{
  println("FilgeLogger")
  val fileOutput=new PrintWriter("file.log")
  fileOutput.println("#")

  def log(msg:String):Unit={
    fileOutput.print(msg)
    fileOutput.flush()
  }
}

new FileLogger {}.log("trait")

}
```

代码执行结果：

```
Logger
FilgeLogger
```

上述代码中定义了两个 trait，trait Logger 中加入了一条执行语句 println("Logger")，trait FileLogger 混入了 Logger 并在定义时加入了 println("FilgeLogger")、val fileOutput=new PrintWriter("file.log") 及 fileOutput.println("#") 3 条执行语句。new FileLogger {}.log("trait") 使用匿名类创建对象，然后调用 log 方法，程序执行完查看 file.log 文件，可以看到文件内容如图 7-1 所示。

图 7-1　file.log 文件的内容

代码 new FileLogger {}.log("trait") 执行时，首先调用 trait Logger 中的构造函数，然后再调用 trait FileLogger 中的构造函数，最后调用匿名类的构造函数从而完成对象的创建。实际构造函数是按以下顺序执行的：

如果有超类，则先调用超类的函数。
如果混入的 trait 有父 trait，它会按照继承层次先调用父 trait 的构造函数。
如果有多个父 trait，则按顺序从左到右执行。
所有父类构造函数和父 trait 被构造完之后，才会构造本类的构造函数。

再来看下面的例子：

```scala
/**
 * trait 构造顺序
 */
object Example07_06 extends App{
  trait Logger{
    println("Logger")
  }
  trait FileLogger extends Logger{
    println("FileLogger")
  }
  trait Closable{
    println("Closable")
  }
  class Person{
    println("Constructing Person....")
  }
  class Student extends Person with FileLogger with Closable{
    println("Constructing Student....")
  }
  new Student
}
```

代码执行结果如下：

```
Constructing Person....
Logger
FileLogger
Closable
Constructing Student....
```

从执行结果中可以看到，程序首先执行的是直接父类 Person 的构造函数，由于混入的第一个 trait FileLogger 扩展自 Logger，因此它会先调用 Logger 的构造函数，然后再调用 FileLogger 的构造函数，完成后再调用 Closable 的构造函数，最后才执行 Student 类自身的构造函数，从而完成对象的创建。

7.2.3 提前定义与懒加载

前面的 FileLogger 中的文件名被写为 "file.log"，程序不具有通用性，这里对 FileLogger 进行改造，增加成员变量 fileName，示例代码如下：

```scala
trait FileLogger extends Logger{
  //增加了抽象成员变量
  val fileName:String
  //将抽象成员变量作为 PrintWriter 参数
```

```
    val fileOutput=new PrintWriter(fileName:String)
    fileOutput.println("#")

    def log(msg:String):Unit={
      fileOutput.print(msg)
      fileOutput.flush()
    }
}
```

这样的话可以使 FileLogger 更加通用，不过这样的设计会存在一个问题，子类虽然可以对 fileName 抽象成员变量进行重写，编译也能通过，但实际执行时会出现空指针异常的问题，请看下列完整示例代码：

```
object Example07_07 extends App{
  import java.io.PrintWriter

  trait Logger{
    def log(msg:String):Unit
  }

  trait FileLogger extends Logger{
    //增加了抽象成员变量
    val fileName:String
    //将抽象成员变量作为 PrintWriter 参数
    val fileOutput=new PrintWriter(fileName:String)
    fileOutput.println("#")

    def log(msg:String):Unit={
      fileOutput.print(msg)
      fileOutput.flush()
    }
  }

  class Person
  class Student extends Person with FileLogger{
    //Student 类对 FileLogger 中的抽象字段进行重写
    val fileName="file.log"
  }
  new Student().log("trait demo")
}
```

代码执行时抛出的异常如下：

```
Exception in thread "main" java.lang.NullPointerException
    at java.io.FileOutputStream.<init>(FileOutputStream.java:212)
    at java.io.FileOutputStream.<init>(FileOutputStream.java:110)
```

```
        at java.io.PrintWriter.<init>(PrintWriter.java:184)
        at cn.scala.chapter07.Example07_07$FileLogger$class.$init$(chapter07.scala:224)
        at cn.scala.chapter07.Example07_07$Student.<init>(chapter07.scala:234)
        at cn.scala.chapter07.Example07_07$delayedInit$body.apply(chapter07.scala:239)
    //其他省略
```

原因就在于构造函数的执行顺序,在创建 Student 对象时,首先调用的是父类 Person 的构造函数,这没有问题,然后再调用 Logger 的构造函数,这也没有问题,但在调用 FileLogger 构造函数时会执行语句 val fileOutput=new PrintWriter(fileName:String)时,由于执行时成员变量 fileName 没有被赋值,从而报空指针异常。

有两种方法可以解决这个问题。

(1) 提前定义

提前定义是指在常规构造之前将变量初始化,例如:

```scala
/**
 * 提前定义
 */
object Example07_08 extends App {
  import java.io.PrintWriter

  trait Logger {
    def log(msg: String): Unit
  }

  trait FileLogger extends Logger {

    val fileName: String
    val fileOutput = new PrintWriter(fileName: String)
    fileOutput.println("#")

    def log(msg: String): Unit = {
      fileOutput.print(msg)
      fileOutput.flush()
    }
  }

  class Person

  class Student extends Person with FileLogger {
    val fileName = "file.log"
  }
```

```
val s = new {
  //提前定义
  override val fileName = "file.log"
} with Student
s.log("predifined variable ")
}
```

上述代码中的

```
val s = new {
  //提前定义
  override val fileName = "file.log"
} with Student
```

为提前定义的使用，在创建对象 Student 对象之前，通过将成员变量 fileName 具体化，从而在创建对象的时候避免出现空指针异常。

```
new {
  //提前定义
  override val fileName = "file.log"
}
```

可以看到，提前定义的方式代码不够优雅，推荐使用下面要讲的懒加载方式。

（2）懒加载

```
/**
 * 懒加载
 */
object Example07_09 extends App {
  import java.io.PrintWriter

  trait Logger {
    def log(msg: String): Unit
  }

  trait FileLogger extends Logger {

    val fileName: String
    //成员变量定义为 lazy
    lazy val fileOutput = new PrintWriter(fileName: String)

    def log(msg: String): Unit = {
      fileOutput.print(msg)
```

```
    fileOutput.flush()
  }
}

class Person

class Student extends Person with FileLogger {
  val fileName = "file.log"
}

val s = new Student
s.log("#")
s.log("lazy demo")
}
```

Example07_09 可以很好地避免创建对象时空指针异常，而且代码也足够优雅。代码 lazy val fileOutput = new PrintWriter(fileName: String)中声明了变量 fileOutput 为 lazy，这样做的好处是创建对象时不会执行这条语句，而是等到真正使用该变量时才会执行，即当执行 s.log("#") 时，由于方法中会执行 fileOutput.print(msg) 和 fileOutput.flush() 两条语句，此时 lazy val fileOutput = new PrintWriter(fileName: String)被真正执行，而此时 Student 对象已经被创建好，fileName 已经被赋值为"file.log"，从而避免了程序运行时的空指针异常问题。

7.3 trait 与类

7.3.1 trait 与类的相似点

trait 可以像普通类一样，定义成员变量和成员方法，而无论其成员变量与成员方法是具体的还是抽象的，trait 在抽象程度上更接近于抽象类，例如：

```
scala> trait Logger{
  println("Logger")
  def log1(msg:String):Unit
  def log2(msg:String):Unit=println(msg)
}
defined trait Logger

scala> abstract class Logger{
  println("Logger")
  def log1(msg:String):Unit
  def log2(msg:String):Unit=println(msg)
}
defined class Logger
```

代码中定义一个 trait Logger 和抽象类 Logger，除定义形式不同外，它们都存在执行语句、抽象方法和具体方法。在使用语法上也有着相似之处，普通的类都可以使用 extends 关键字扩展类或混入 trait，例如：

```scala
//Logger 为 trait
scala> trait Logger{
  println("Logger")
    def log1(msg:String):Unit
    def log2(msg:String):Unit=println(msg)
}
defined trait Logger

//使用 extends 关键字混入 Logger
scala> class Person extends Logger{
    def log1(msg:String):Unit=println("log1:"+msg)
}
defined class Person

scala> val p=new Person
Logger
p: Person = Person@713caa

scala> p.log1("Person extends Logger trait")
log1:Person extends Logger trait
```

上面的代码给出的是 Logger 为 trait，Person 使用 extends 关键字混入 Logger，然后实现混入 trait 中的抽象方法。对于抽象类，普通类同样使用 extends 关键字实现对该抽象类的继承，例如：

```scala
//将 Logger 定义为抽象类
scala> abstract class Logger{
  println("Logger")
    def log1(msg:String):Unit
    def log2(msg:String):Unit=println(msg)
}
defined class Logger

//通过 extends 关键字扩展类 Logger，实现继承
scala> class Person extends Logger{
    def log1(msg:String):Unit=println("log1:"+msg)
}
defined class Person

scala> val p=new Person
Logger
```

```
p: Person = Person@15a9d03

scala> p.log1("Person extends abstract class Person")
log1:Person extends abstract class Person
```

通过代码不难发现，除了将 Logger 声明为抽象类之外，和将 Logger 定义为 trait 时的代码是相同的。只不过从现实角度来看，Person 类作为 Logger 类的子类，在认知上不符合逻辑，因为它们之间不构成"is-a"的关系，而 Person 类混入 Logger trait，更符合我们的认知习惯。

还有一个值得注意的地方就是在定义 trait 时可以使用 extends 关键字继承类，例如：

```
//定义一个普通类
scala> class A{
  val msg:String="msg"
}
defined class A

//trait B 继承自类 A
scala> trait B extends A{
  def print()=println(msg)
}
defined trait B

scala> new B{}.print()
msg
```

7.3.1 trait 与类的不同点

trait 与类之间还存在许多的不同之处，首先无论是普通的类还是抽象类都可以在类定义时使用主构造函数定义类的成员变量，但 trait 不能，例如：

```
scala> abstract class Logger(val msg:String)
defined class Logger

scala> trait Logger(val msg:String)
<console>:1: error: traits or objects may not have parameters
    trait Logger(val msg:String)
                ^
```

通过上述代码可以看到，trait 不能有使用主构造函数定义的成员变量，这是 trait 与类间的一个重要区别。

其次，Scala 语言中的类不能继承多个类，但可以混入多个 trait，例如：

```
scala> trait A
defined trait A


scala> trait B
```

```
defined trait B

//类可以混入多个 trait
scala> class C extends A with B
defined class C
```

7.4 多重继承问题

Scala 语言中可以通过使用 trait 实现多重继承，不过在实际使用时常常会遇到菱形继承的问题，如以下代码：

```
scala> trait A{
  def print:Unit
}
defined trait A

scala> trait B1 extends A{
  var B1="Trait B1"
  override def print=println(B1)
}
defined trait B1

scala> trait B2 extends A{
  var B2="Trait B2"
  override def print=println(B2)
}
defined trait B2

scala> class C extends B1 with B2
defined class C

scala> val c=new C
c: C = C@cd48fc

//使用的是 trait B2 中的 print 方法
scala> c.print
Trait B2
```

上述代码存在菱形继承问题，即 trait B1、trait B2 中分别混入了 trait A，然后类 C 又混入了 B1、B2，这导致类 C 中会存在两个 print 方法。为解决方法调用时的冲突问题，Scala 会对类进行线性化，在存在多重继承时会使用最右深度优先遍历算法查找调用的方法，例如 class C extends B1 with B2 混入了 trait B1 和 B2，在调用 print 方法时会采用最右深度优先遍历算法查找，在本例中为 B2 中的 print 方法，而 B1 中的 print 方法没有被执行。

```
scala> trait A{
  val a="Trait A"
  def print(msg:String)=println(msg+":"+a)
}
defined trait A

scala> trait B1 extends A{
  val b1="Trait B1"
  override def print(msg:String)=super.print(msg+":"+b1)
}
defined trait B1

scala> trait B2 extends A{
  val b2="Trait B2"
  override def print(msg:String)=super.print(msg+":"+b2)
}
defined trait B2

scala> class C extends B1 with B2
defined class C

scala> new C().print("print method In")
print method In:Trait B2:Trait B1:Trait A
```

在上述代码中，在 trait A 中定义了方法 def print(msg:String)=println(msg)，然后在 trait B1、B2 中对方法进行重写，只不过在方法实现中使用 super 关键字进行父类的方法调用，如 super.print(msg+":"+b1)、super.print(msg+":"+b2)。从代码的执行结果可以看到，new C().print("print method In")输出内容为 print method In:Trait B2:Trait B1:Trait A。为什么结果输出是这样的呢？事实上这种 super 关键字的使用方式也是一种惰性求值，super 关键字调用的方法不会马上执行而是在真正被调用时执行，它的执行原理同样按照最右深度优先遍历算法进行，先将 B2、B1、A 中的成员变量按序组装得到:Trait B2:Trait B1:Trait A，然后再调用 print("print method In"+":Trait B2:Trait B1:Trait A")得到最终结果，这种方式是解决多重继承菱形问题的最常用方法。

7.5　自身类型

Scala 官方对自身类型的定义是："A self type of a trait is the assumed type of this, the receiver, to be used within the trait. Any concrete class that mixes in the trait must ensure that its type conforms to the trait's self type. The most common use of self types is for dividing a large class into several traits"[1]。具体来说，任何混入该特质的具体类必须确保它的类型符合特质的自身类型，自身类型最通常的应用是将大类分成若干特质。下面的代码定义了特质 A 的自身类型：

```
scala> trait B
defined trait B

//通过 this:B=>定义自身类型
scala> trait A { this: B => }
defined trait A

//在使用时必须混入特质 B，否则会报错
scala> class C extends A
<console>:9: error: illegal inheritance;
 self-type C does not conform to A's selftype A with B
       class C extends A

scala> class C extends A with B
defined class C
```

为什么说自身类型最通常的应用是将大类分为若干特质呢？先看下面这个特质[2]

```
scala> :paste
// Entering paste mode (ctrl-D to finish)

trait Prompter1 {

    val prompt = "> "
    val greeting = "Hello world"

    def printGreeting() {
        println(prompt + greeting)
    }
}

// Exiting paste mode, now interpreting.
defined trait Prompter1

scala> val prompter1 = new Object with Prompter1
prompter1: Prompter1 = $anon$1@d99324

scala> prompter1.printGreeting
```

---

[1] http://docs.scala-lang.org/glossary/#self-type

[2] http://marcus-christie.blogspot.my/2014/03/scala-understanding-self-type.html

> Hello world

通过自身类型可以将 Prompter1 进行拆分，具体代码如下：

```
scala> :paste
// Entering paste mode (ctrl-D to finish)

trait GreetingProvider {
   val greeting = "Hello world"
}

// Exiting paste mode, now interpreting.

defined trait GreetingProvider

scala>

scala> :paste
// Entering paste mode (ctrl-D to finish)

trait Prompter2 {
   // 自身类型标记
   this: GreetingProvider =>

   val prompt = "> "

   def printGreeting() {
     //与 Prompter1 中的代码一样
     println(prompt + greeting)

     //下面的代码也是可行的, this.prompt 中的 this 为 Prompter2 对应的对象
     //而 this.greeting 中的 this 为 GreetingProvider 对应的对象
     println(this.prompt + this.greeting)
   }
}
// Exiting paste mode, now interpreting.

defined trait Prompter2

scala> val prompter2 = new Prompter2 with GreetingProvider
prompter2: Prompter2 with GreetingProvider = $anon$1@5a6f01

scala> prompter2.printGreeting
> Hello world
> Hello world
```

可以看到，Prompter1 通过自身类型被拆分为两个 trait，分别是 Prompter2 和 GreetingProvider，更复杂的例子见链接[1]。

自身类型还有用于别名的一种特殊用途，示例代码如下：

```
scala> class A{
  //下面 self =>  定义了 this 的别名，它是 self type 的一种特殊形式
  //这里的 self 并不是关键字，可以是任何名称
  self =>
  val x=2
  //可以用 self.x 作为 this.x 使用
  def foo = self.x + this.x
}
```

上面的代码等价于下列代码：

```
scala> class A{
  //下面 self =>  定义了 this 的别名，它是 self type 的一种特殊形式
  //这里的 self 并不是关键字，可以是任何名称
  self:A=>
  val x=2
  //可以用 self.x 作为 this.x 使用
  def foo = self.x + this.x
}
defined class A
```

这种用法在程序设计中十分常见，例如：

```
class OuterClass {
    outer => //定义了一个外部类别名
    val v1 = "here"
    class InnerClass {
       // 用 outer 表示外部类，相当于 OuterClass.this
       println(outer.v1)
    }
}
```

小 结

本章详细介绍了 Scala 中的 trait 的重要特性及其常见使用方法，trait 虽然与 Java 语言中的 interface、抽象类有着诸多的相似之处，但 trait 有着更为强大和灵活的使用方式，它完全可以替代 Java 语言中的 interface、抽象类和 Scala 中的抽象类来实现相同的功能。本章也重点介绍了在混入 trait 时的菱形继承问题及如何解决多个方法的冲突问题，同时也对面向对象编程中的其他特性做了介绍。在下一章中，将介绍 Scala 语言中的代码组织及访问控制利器——包（package）。

---

1 http://marcus-christie.blogspot.my/2014/03/scala-understanding-self-type.html

第 8 章　包（package）

　　Java 语言中使用包来进行大型工程代码的组织，在 Scala 语言中也是如此，Scala 包与 Java 包有着诸多相似之处，但 Scala 包提供了更灵活的访问控制、定义与使用方式。本章将对这部分相关内容进行详细介绍。

8.1 包的定义

同 Java 中的包、C++中的命名空间一样，Scala 中的包主要用于大型工程代码的组织，同时也解决命名冲突的问题。Scala 中的包与 Java 有着诸多相似之处，但 Scala 语言中的包更加灵活。

下面的代码给出了 Scala 包的定义：

```scala
//将代码组织到 cn.scala.chapter08 包中
package cn.scala.chapter08

abstract class Animal {
  //抽象字段(域)
  var height:Int
  //抽象方法
  def eat:Unit
}
class Person(var height:Int) extends Animal{
  override def eat()={
    println("eat by mouth")
  }
}
object Person extends App{
  new Person(170).eat()
}
```

图 8-1 中左边是代码文件结构图，右边是编译后生成的字节码文件结构图。

图 8-1　左为代码文件结构图，右为编译后生成的字节码文件结构图

可以看出，Scala 包的这种组织方式与 Java 包的组织方式没有明显区别。不过 Scala 包组织方式可以更灵活，例如：

```scala
//下面的代码定义了一个 cn.scala.chapter08，在该包中定义一个类 Teacher
//在程序的任何地方都可以通过 cn.scala.chapter08.Teacher 来使用 Teacher 这个类
package cn{
  package scala{
    package chapter08{
      class Teacher {
```

```
            }
          }
        }
      }
```

上述代码可以在 Scala 程序中的任何地方出现，它表示在 cn.scala.chapter08 包中定义了一个类 Teacher，定义完成后可以通过 cn.scala.chapter08.Teacher 在程序中其他任何地方使用该类。上述代码保存为 Teacher.scala 文件，放在 Scala 源文件的根目录下，如图 8-2 左边所示，编译完成后如图 8-2 右边所示。

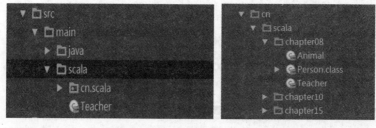

图 8-2　左为 Scala 源文件的根目录，右为编译完成后的后的文件结构图

从上面的代码不难看出，Scala 可以灵活定义包并在该包中定义程序所需的类，帮我们自动组织生成的字节码文件。将上述 Teacher.scala 文件内容修改如下：

```
//下面的代码定义了一个 cn.scala.chapter08
//在程序的任何地方都可以通过 cn.scala.chapter08.Teacher 来使用 Teacher 这个类
package cn{
  package scala{
    package chapter08{
      class Teacher {


      }
    }
  }
}

package cn{
  package scala{
    package spark{
      class SparkDemo {


      }
    }
  }
}
```

编译后生成的字节码文件结构如图 8-3 所示。

图 8-3 字节码文件结构

相信通过前面的例子我们已经能够掌握如何定义 Scala 包及理解 Scala 包是如何进行代码组织的。虽然 Scala 包的定义与代码组织非常灵活，在程序的任何地方都可以新定义包及在该包中定义程序需要的类，但这种方式会使代码非常分散，不便于进行代码集中管理，在实际使用时还是推荐使用类似 Java 包的组织方式。

8.2　包的使用和作用域

8.2.1　包的使用

下面的示例演示了包的使用及访问规则。

```
package cn{
  package scala{
    //在包 cn.scala 下创建了一个 Utils 单例
    object Utils{
      def toString(x:String){
        println(x)
      }
      //外层包无法直接访问内层包，下面这一行代码编译通不过
      //def getTeacher():Teacher=new Teacher("john")
      //如果一定要使用的话，可以引入包
      import cn.scala.chapter08._
      def getTeacher():Teacher=new Teacher("john")
    }
    //定义了 cn.scala.chapter08
    package chapter08{
      class Teacher(var name:String) {
        //演示包的访问规则
        //内层包可以访问外层包中定义的类或对象，无需引入
        def printName()={Utils.toString(name)}
      }
    }
  }
}
```

```
object AppDemo{
  //scala 允许在任何地方进行包的引入，_的意思是引入该包下的所有类和对象
  import cn.scala._
  import cn.scala.chapter08._
  def main(args: Array[String]): Unit = {
    Utils.toString(new Teacher("john").name)
    new Teacher("john").printName()
  }
}
```

在上述代码中定义了包 cn.scala，然后该包中定义了一个单例对象 Utils 和子包 chapter08，在子包 chapter08 中又定义了类 Teacher。在单例对象 Utils 中直接定义 def getTeacher():Teacher=new Teacher("john")方法会出错，这是因为外层包感知不到内层包的存在，必须使用 import cn.scala.chapter08._ 将 chapter08 包中的类引入到当前作用域才行，而 Teacher 类中方法 def printName()={Utils.toString(name)}在没有引入 Utils 单例对象的情况下可以直接使用，这是因为内层包中的类可以感知到外层包中类的存在。应用程序对象 AppDemo 在使用 Utils 和 Teacher 时必须通过 import cn.scala._ 和 import cn.scala.chapter08._ 将它们引入到当前作用域才能够通过编译。

8.2.2 包作用域

在 Java 语言中，主要通过 public、private、protected 及默认控制来实现包中类成员的访问控制，当定义一个类时，如果类成员不加任何访问控制符时，表示该类成员在定义该类的包中可见。在 Scala 中没有 public 关键字，仅有 private 和 protected 访问控制符，当一个类成员不加 private 和 protected 时，它的访问权限就是 public。在 Scala 中提供了更为灵活的访问控制方法，private、protected 除了可以直接修饰成员外，还可以以 private[X]、protected[X]的方式进行更为灵活的访问控制，这种访问控制的意思是可以将 private、protected 限定到 X，X 可以是包、类，还可以是单例对象。下面以 private 关键字的包访问控制为例，说明包的作用域问题。

在第 6 章成员访问控制部分已经对 private 的访问控制规则进行了介绍，我们知道 private 关键字修饰的成员在类及伴生对象中都可以访问。示例如下：

```
package cn.scala.chapter8;

class Teacher(var name: String) {
  private def printName(tName:String="") :Unit= { println(tName) }
  //可以访问
  def print(n:String)=this.printName(n)
}

object Teacher{
  //伴生对象可以访问
  def printName=new Teacher("john").printName()
```

```
}
object appDemo {
  def main(args: Array[String]): Unit = {
      //不能访问
      //new Teacher("john").printName()
  }
}
```

接下来,给出的是 private[包名]限定的访问控制示例。

```
package cn{
  class UtilsTest{
    //编译通不过,因为Utils利用private[scala]修饰,只能在scala及其子包中使用
    //Utils.toString()
  }
  package scala{
    //private[scala]限定 Utils 只能在 scala 及子包中使用
    private[scala] object Utils{
      def toString(x:String){
        println(x)
      }
      import cn.scala.chapter8._
      def getTeacher():Teacher=new Teacher("john")

    }
    package chapter8{
      class Teacher(var name:String) {
          def printName()={Utils.toString(name)}
      }

    }
  }
}
object appDemo{
    import cn.scala._
    import cn.scala.chapter8._
    def main(args: Array[String]): Unit = {
        //编译通不过,同UtilsTest
        //Utils.toString(new Teacher("john").name)
        new Teacher("john").printName()
    }
}
```

上述代码中,包 cn 中定义了类 UtilsTest 及子包 scala,包 cn.scala 中又定义了单例对象 Utils 及子包 chapter8,而在包 cn.scala.chapter8 中又定义了类 Teacher。使用时需要注意,类 UtilsTest 和应用程序对象 appDemo 中不能访问单例对象 Utils,因为 object Utils 定义时使用了 private[scala]限定,它的意思是单例对象 Utils 只能在 Scala 包及子包中访问,在其他地方不能访问,示例中类 cn.scala.chapter8. Teacher 中的 def printName()={Utils.toString(name)}可以直接

使用 Utils.toString(name)，便是这个原因。

最后，我们再介绍 private[this]的使用，private[this]限定只有该类的对象才能访问，即使是伴生对象也不能访问，private[this]修饰的成员也被称为对象私有成员，具体示例如下：

```scala
package cn.scala.chapter8;

class Teacher(var name: String) {
  private[this] def printName(tName:String="") :Unit= { println(tName) }
  //调用 private[this] printName 方法
  def print(n:String)=this.printName(n)
}

object Teacher{
   //private[this]限定的成员，即使伴生对象 Teacher 也不能使用
  //def printName=new Teacher("john").printName()
}

object AppDemo {
  def main(args: Array[String]): Unit = {
      //编译不能通过
      //new Teacher("john").printName()
  }
}
```

代码中 cn.scala.chapter8 包中分别定义了伴生类 Teacher、伴生对象 Teacher 及应用程序对象 AppDemo，伴生类 Teacher 中使用 private[this] def printName(tName:String="") :Unit= { println(tName) }定义了成员方法 printName，该方法只能在伴生类 Teacher 内部访问如 def print(n:String)=this.printName(n)，但不能在伴生对象 Teacher 和其他外部如 AppDemo 中访问。也就是说 private[this]对访问进行了严格限定，使用 private[this]修饰的成员只能在本类中使用，在外部不能访问，即使是伴生对象也不能访问。

protected[X]修饰的成员其访问规则与 private[X]类似，不同的是 protected 可以影响到其子类。包访问规则表如表 8-1 所示。

表 8-1 包访问规则表

修饰符	访问范围
无任何修饰符	任何地方都可以使用
private[scala]	在定义的类中可以访问，在 Scala 包及子包中可以访问
private[this]	只能在定义的类中访问，即使伴生对象也不能访问
private	在定义的类及伴生对象中可以访问，其他地方不能访问
protected[scala]	在定义的类及子类中可以访问，在 Scala 包及子包中可以访问
protected[this]	只能在定义的类及子类中访问，即使伴生对象也不能访问
protected	在定义的类及子类中访问，伴生对象可以访问，其他地方不能访问

8.3 包 对 象

我们知道，Scala 语言是纯面向对象编程语言，语言中的一切皆对象，包也不例外，既然是对象，自然它就可以拥有属性和方法，通过 package 关键字声明的对象被称为包对象（package object）。包对象主要用于对常量和工具函数的封装，使用时直接通过包名引用。

```
//下面的代码给出了包对象的定义
package cn.scala.chapter8

//利用package关键字定义单例对象
package object Math {
  val PI=3.141529
  val THETA=2.0
  val SIGMA=1.9
}

class Computation{
  def computeArea(r:Double)=Math.PI*r*r
}
```

将上述代码保存为 package.scala 文件，代码组织结构如图 8-4~图 8-6 所示。

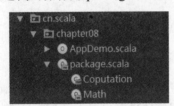
图 8-4　package.scala 代码组织结构

图 8-5　编译后的 package.scala 代码组织结构

图 8-6　package.scala 代码组织结构

将 package$.class 和 package.class 字节码文件进行反编译，得到如下代码。

```
E:\IntellijIDEAWorkspace\out\production\ScalaProject\cn\scala\chapter08\Math>javap -private package$.class
Compiled from "package.scala"
public final class cn.scala.chapter08.Math.package$ {
  public static final cn.scala.chapter08.Math.package$ MODULE$;
  private final double PI;
```

```
    private final double THETA;
    private final double SIGMA;
    public static {};
    public double PI();
    public double THETA();
    public double SIGMA();
    private cn.scala.chapter08.Math.package$();
}

E:\IntellijIDEAWorkspace\out\production\ScalaProject\cn\scala\chapter08\Math>jav
ap -private package.class
Compiled from "package.scala"
public final class cn.scala.chapter08.Math.package {
    public static double SIGMA();
    public static double THETA();
    public static double PI();
}
```

不难看出,Scala 为包对象 Math 创建了一个文件夹,然后创建了两个类,通过单例的方式实现方法调用。

8.4 import 高级特性

Java 语言中的 import 关键字所充当的语法作用是将当前程序所需要的类引入,Scala 语言中的 import 关键字除了拥有将类或对象引入到当前作用域的功能之外,还有一些在实际中应用非常广泛的其他语法特性,如引入重命名、类隐藏等功能。

8.4.1 隐式引入

如果不引入任何包,Scala 会默认引入 java.lang._ 和 scala.Predef.对象中的所有类和方法,在 Scala REPL 命令行中可以输入:import 命令查看默认的引入,具体代码如下:

```
scala> :import
 1) import scala.Predef._          (162 terms, 78 are implicit)
```

可以看到,Scala 会自动引入 import scala.Predef._,即将定义在 Predef 对象中的所有类、成员变量和方法引入到当前作用域。由于 Scala 默认会自动帮我们引入,所以也称为隐式引入。

8.4.2 引入重命名

Scala 中允许对引入的类或方法进行重命名,例如需要在程序中同时使用 java.util.HashMap 及

scala.collection.mutable.HashMap 时，可以利用引入重命名消除命名冲突的问题（虽然也可以采用包名前缀的方式使用，但代码不够简洁）。示例如下：

```scala
//将java.util.HashMap重命名为JavaHashMap
import java.util.{ HashMap => JavaHashMap }
import scala.collection.mutable.HashMap
object RenameUsage {
  def main(args: Array[String]): Unit = {
    val javaHashMap = new JavaHashMap[String, String]()
    javaHashMap.put("Spark", "excellent")
    javaHashMap.put("MapReduce", "good")
    for(key <- javaHashMap.keySet().toArray){
      println(key+":"+javaHashMap.get(key))
    }

    val scalaHashMap=new HashMap[String,String]
    scalaHashMap.put("Spark", "excellent")
    scalaHashMap.put("MapReduce", "good")
    scalaHashMap.foreach(e=>{
      val (k,v)=e
      println(k+":"+v)
    })
  }
}
```

代码中同时引入了 java.util.HashMap 和 scala.collection.mutable.HashMap，如果不采用引入重命名会产生名称冲突，通过 java.util.{ HashMap => JavaHashMap }将java.util.HashMap 重命名为 JavaHashMap 从而达到消除冲突的目的，在使用时直接使用 new JavaHashMap[String, String]() 创建 java.util.HashMap 对象，通过使用 new HashMap[String,String] 创建 scala.collection.mutable.HashMap 对象。

8.4.3 类隐藏

在 Scala 程序中，有时候不希望引入某个包中的若干个类，例如代码中需要使用类 scala.collection.mutable.HashMap 和 java.util 包中除 HashMap 外的所有类，即希望避免使用 java.util.HashMap 及它所带来的命名冲突问题。Scala 提供了类隐藏机制来解决这一问题，具体示例如下：

```scala
//通过HashMap=> _，这样类便被隐藏起来了
import java.util.{HashMap=> _,_}
import scala.collection.mutable.HashMap
```

```
object ClassHiddenUsage {
  def main(args: Array[String]): Unit = {

    //这样的话,HashMap便无歧义地指向scala.collection.mutable.HashMap
    val scalaHashMap=new HashMap[String,String]
    scalaHashMap.put("Spark", "excellent")
    scalaHashMap.put("MapReduce", "good")
    scalaHashMap.foreach(e=>{
      val (k,v)=e
      println(k+":"+v)
    })
  }
```

代码 import java.util.{HashMap=> _,_} 的意思是引入 java.util 包中所有类的同时将 HashMap 隐藏。这样 java.util.HashMap 便不会出现在当前作用域，从而避免了与 scala.collection.mutable.HashMap 产生命名冲突，代码中使用的 HashMap 便无歧义地指向 scala.collection.mutable.HashMap。

小　结

本章详细地介绍了 Scala 中包的定义与使用，Scala 包可以像 Java 中的包那样使用，但 Scala 包提供了更为灵活与强大的使用方法。同时，对 Scala 包的访问规则与作用域等相关内容进行了详细介绍，然后对包对象、import 高级使用特性进行了介绍，重点介绍引入时的重命名与类隐藏等内容。在下一章中，将重点介绍 Scala 语言学习最重要的语法内容之一——模式匹配。

第 9 章 模式匹配

　　模式匹配是学习 Scala 语言必备的语法技能之一,它灵活的程序结构更符合人脑的思维处理方式,是使用 Scala 语言进行实际编程的利器。本章将详细介绍 Scala 语言中的模式匹配、模式匹配的 7 大类型、模式匹配原理、正则表达式与模式匹配、for 循环中的模式匹配及模式匹配与样例类、样例对象。

9.1 模式匹配简介

模式匹配是 Scala 提供的最重要的语言特性，在实际开发过程中应用非常广泛，可以说模式匹配在 Scala 语言中无处不在，如下列赋值代码所示。

```
scala> val (first,second)=(1,2)
first: Int = 1
second: Int = 2
```

在给变量赋值时，上面的赋值语法经常会被使用，但实际上其背后的原理便是模式匹配。为了弄清楚什么是模式匹配，先来看一下 Java 语言的 switch 语句。

```java
package cn.scala.chapter09;
/**
 * Java Switch 语句使用演示
 */
public class JavaExample9_1 {
    public static void main(String[] args) {
        for (int i = 1; i < 5 ; i++) {
            switch (i){
                //带 break 语句
                case 1: System.out.println("1");
                    break;
                //不带 break 语句，会意外陷入其他分支，同时输出 2、3
                case 2: System.out.println("2");
                case 3: System.out.println("3");break;
                default:System.out.println("default");
                //case 语句后不能接表达式
                //case i%5==0: System.out.println("10");
            }
        }
    }
}
```

代码运行结果如下：

```
1
2
3
3
default
```

在 Java 语言中，case 语句后需要加上 break 语句，如 case 1: System.out.println("1");break;，如果没有该 break 语句，则程序会意外陷入分支，如 case 2: System.out.println("2");后没有 break 语句的话，执行完该语句后还会接着执行 case 3: System.out.println("3");break;，这可能与实际

程序运行要求相去甚远，而 Java 开发人员在实际编程过程中很容易忽略 break 语句。Scala 语言的模式匹配可以解决这一问题，例如：

```
package cn.scala.chapter09
/**
 * 使用Scala中的模式匹配可以避免Java语言中switch语句会意外陷入分支的情况
 *
 */
object ScalaExampel9_1 extends App{
for(i<- 1 to 5)
  //Scala模式匹配
  i match {
    //仅匹配值为1的情况，不会意外陷入分支，输出1
    case 1=> println(1)
    case 2=> println(2)
    case 3=> println(3)
    //_通配符表示，匹配其他情况，与Java switch语句中的default相同
    case _=>println("其他")
  }
}
```

示例运行结果如下：

```
1
2
3
其他
其他
```

可以看到，首先 Scala 语言中的模式匹配可以有效地避免 Java switch 语句意外陷入另外一个分支的情况。其次，Scala 模式匹配代码更简洁，更符合人脑的思维方式。第三，Scala 语言中的模式匹配比 Java 语言更强大的地方在于模式匹配中的 case 语句还可以加入表达式，例如：

```
/**
 * Scala模式匹配中的case语句还可以加入表达式
 *
 */
object ScalaExampel9_2 extends App{
  for(i<- 1 to 6)
  //Scala模式匹配
    i match {
      case 1=> println(1)
      //case 语句中可以加入表达式
      case x if (x%2==0)=>println(s"$x 能够被2整除")
      //其他情况则不进行任何操作
```

```
      case _ =>
    }
}
```

代码运行结果如下：

```
1
2 能够被 2 整除
4 能够被 2 整除
6 能够被 2 整除
```

可以看到，case x if (x%2==0)=>println(s"$x 能够被 2 整除")语句能够匹配所有能够被 2 整除的整数，case 语句跟的是一个变量 x，然后紧接着加入表达式 if (x%2==0)判断变量 x 能否被 2 整除，如果能的话，则执行后面的语句。

Scala 模式匹配除了可以在一般程序中使用外，还可以作为函数体，模式匹配结果作为函数返回值，使用方法如下：

```
/**
  * 函数中使用模式匹配，模式匹配结果作为函数返回值
  *
  */
object ScalaExampel9_3 extends App{
  // patternMatching 函数中使用模式匹配，匹配结果作为函数的返回值
  def patternMatching(x:Int)=x match {
    case 5 => "整数 5"
    case x if(x%2==0) =>"能被 2 整除的数"
    case _ =>"其他整数"
  }
  println(patternMatching(5))
  println(patternMatching(4))
  println(patternMatching(3))
}
```

代码运行结果如下：

```
整数 5
能被 2 整除的数
其他整数
```

可以看到，模式匹配作为函数 patternMatching 的函数体，模式匹配结果为函数的返回值，这种使用方式在实际应用中非常普遍。

9.2 模式匹配的 7 大类型

9.2.1 常量模式

常量模式指的是 case 语句后面接的全部为常量，如 ScalaExampel9_1 中的代码片段：

```
for(i<- 1 to 5)
   i match {
      case 1=> println(1)
      case 2=> println(2)
      case 3=> println(3)
      case _ =>println("其他")
   }
```

上述代码中使用的便是常量模式，当然 Scala 语言中还存在其他类型的常量，如 null、Nil（空列表）、布尔常量（true 和 false）及字符型常量等，它们都可以在常量模式中使用。

9.2.2 变量模式

变量模式指的是模式匹配中的 case 语句后面接的是变量，如 ScalaExampel9_2 中的代码片段：

```
for(i<- 1 to 6)
   i match {
      //常量模式
      case 1=> println(1)
      //变量模式
      case x if (x%2==0)=>println(s"$x 能够被 2 整除")
      case _ =>
   }
}
```

代码 case x if (x%2==0)=>println(s"$x 能够被 2 整除")使用的便是变量模式，变量模式可以匹配任何值，它可以与条件判断表达式（也称为守卫）结合起来使用，从而灵活控制程序的行为。例如：

```
/**
 * 变量模式
 */
object ScalaExampel9_4 extends App{
   def patternMatching(x:Int)=x match {
      case i if(i%2==0) =>"能被 2 整除的数"
      case i if(i%3==0) =>"能被 3 整除的数"
      case i => "除能被 2 或 3 整除外的其他整数"
```

```
    }
    println(patternMatching(5))
    println(patternMatching(4))
    println(patternMatching(3))
}
```

代码运行结果如下:

除能被 2 或 3 整除外的其他整数
能被 2 整除的数
能被 3 整除的数

代码中的 case i if(i%2==0) =>"能被 2 整除的数"便是变量模式匹配的应用,变量 i 会匹配任何值但由于后面有守卫条件 if(i%2==0),所以它只会匹配能够被 2 整除的整数。同理,语句 case i if(i%3==0) =>"能被 3 整除的数"也是变量模式,只会匹配能够被 3 整除的整数。需要注意的是,语句 case i => "除能被 2 或 3 整除外的其他整数",由于该语句是放在最后的,它会匹配所有不能够被 2 或 3 整除的整数。如果 case i =>这条语句放在 case i if(i%2==0)=>及 case i if(i%3==0)这两条语句之前,则该语句后面的代码将不会被执行,具体示例如下:

```
/**
 * 注意变量模式匹配顺序
 *   case i =>这种变量匹配模式的作用等同 case _
 */
object ScalaExampel9_5 extends  App{
  def patternMatching(x:Int)=x match {
    //case i 变量模式等同于 case _
    case i => "匹配任何整数,后面两条变量模式不会执行"
    case i if(i%2==0) =>"能被 2 整除的数"
    case i if(i%3==0) =>"能被 3 整除的数"
  }
  println(patternMatching(5))
  println(patternMatching(4))
  println(patternMatching(3))
}
```

程序运行结果如下:

匹配任何整数,后面两条变量模式不会执行
匹配任何整数,后面两条变量模式不会执行
匹配任何整数,后面两条变量模式不会执行

从上面的代码中不难看出,变量模式 case i 与 case i if(i%2==0)、case i if(i%3==0)的位置对执行结果是有影响的,因为 case i 会匹配任何值,它与 case _这种通配符的作用是一样的。在实际应用时需要特别留意,像 case i if(i%2==0)和 case i if(i%3==0)这两种变量匹配模式也需要注意其执行顺序,例如 println(patternMatching(6))这条语句,由于 6 既能被 2 整除也能被 3 整除,哪条语句放在前面便会先执行哪一条。

9.2.3 构造函数模式

在 Scala 语言中，代码 class Dog(val name:String, val age:Int)除了定义一个类 Dog 外，还为 Dog 类自动创建了成员域 name 和 age、成员方法及构造函数。通过 val dog=new Dog("Pet",2)调用构造函数创建 Dog 对象，本节介绍的构造函数模式要做的事情与创建对象相反，用于解构对象，具体见下列代码：

```
/**
 * 构造函数模式
 */
object ScalaExampel9_6 extends App{
  //定义一个 case class
  case class Dog(val name:String,val age:Int)
  //利用构造函数创建对象
  val dog=Dog("Pet",2)

  def patternMatching(x:AnyRef)=x match {
    //构造函数模式,其作用与相反,用于对象进行解构（也称析构）
    case Dog(name,age) => println(s"Dog name=$name,age=$age")
    case _ =>
  }
  patternMatching(dog)
}
```

代码运行结果如下：

```
Dog name=Pet,age=2
```

代码中的 val dog=Dog("Pet",2)调用的是 apply 方法创建对象，apply 方法会调用 Dog 的构造函数创建对象，代码 case Dog(name,age) => println(s"Dog name=$name,age=$age")为构造函数模式，该行代码的作用与代码 val dog=Dog("Pet",2)的作用相反，是用于对创建的对象进行解构（也称析构），构造函数模式匹配成功的话，会将对象 dog 中的成员域 name 的值（这里为 "Pet"）、age 的值（这里为 2）分别赋值给模式 case Dog(name,age)中的 name 和 age，从而完成对象成员域的析取。

有时候在使用构造函数模式时，只想析取对象的部分成员域，这时可以用通配符匹配，只给出需要析取的成员域对应的变量即可。

```
/**
 * 构造函数,构取对象的部分成员域
 */
object ScalaExampel9_7 extends App{
  //定义一个 case class
  case class Dog(val name:String,val age:Int)
  //利用构造函数创建对象
  val dog=Dog("Pet",2)
```

```
    def patternMatching(x:AnyRef)=x match {
      //构造函数模式,通配符匹配对象成员域name,但程序中不需要该成员域的值
      case Dog(_,age) => println(s"Dog age=$age")
      case _ =>
    }
    patternMatching(dog)
}
```

代码运行结果如下:

```
Dog age=2
```

代码中的 case Dog(_,age) => println(s"Dog age=$age")的通配符_表示该构造函数模式匹配 x 中的成员域 name 但程序中并不需要使用 name 的值（即匹配但丢弃该值），age 则表示匹配 x 中的成员域 age，同时程序中需要使用 age 的值，因此将其赋值给变量 age。

9.2.4 序列模式

序列模式用于匹配 Seq[+A] 类及子类（如 scala.collection.immutable.Seq、scala.collection.mutable.Seq 等）集合的内容，例如:

```
/**
 * 序列模式
 */
object ScalaExampel9_7 extends App{
  val arrInt=Array(1,2,3,4)
  def patternMatching(x:AnyRef)=x match {
    //序列模式,与构建序列的作用相反,用于对序列中的元素内容进行析取
    case Array(first,second) => println(s"序列中第一个元素=$first,第二个元素=$second")
    //first,second 分别匹配序列中的第一、二个元素, _*匹配序列中剩余其他元素
    case Array(first,_,three,_*) => println(s"序列中第一个元素=$first,第三个元素=$three")
    case _ =>
  }
  patternMatching(arrInt)
}
```

代码运行结果如下:

```
序列中第一个元素=1,第三个元素=3
```

代码 val arrInt=Array(1,2,3,4)创建了一个数组，case Array(first,second)为序列模式，但该模式不能匹配输入的数组 arrInt，因为该数组中有 4 个元素，而 case Array(first,second)只会匹配只有两个元素的数组。case Array(first,_,three,_*)可以匹配输入的数组 arrInt，first 会匹配数组的第一个元素，通配符_会匹配数组的第二个元素，three 会匹配数组的第三个元素，_*则

会匹配数组中的剩余其他若干元素，因此匹配成功后 first 被赋值为 1，three 被赋值为 3。事实上，case Array(first,_,three,_*)会匹配任何数组元素个数超过 4 个的数组。

通过代码可以看到，序列模式与构造函数模式有一定的相似之处，首先都是通过类名来进行匹配，其次都可以使用通配符"_"来匹配不需要析取的内容，但与构造函数模式所不同的是，由于数组中的元素个数可以非常大，序列模式中可以使用"_*"来匹配序列中剩余的若干个元素内容，构造函数模式则不能使用这种方式。

使用序列模式还有个值得注意的地方，那就是"_*"只能放在模式的最后，即只有 case Array(first,_,three,_*)这种形式是合法的，其他如 case Array(first,_*,three)等任何"_*"不放在模式最后的形式都是不合法的。

9.2.5 元组模式

元组在 Scala 语言中是一种十分重要的数据结构，它可以将不同类型的值组合成一个对象，在实际应用中十分广泛。同样在 Scala 模式匹配的应用中，元组模式也占据着十分重要的地位。在讲解元组模式匹配之前，我们选来看一个定义元组的例子：

```
scala> val tuple=("nest",1)
tuple: (java.lang.String, Int) = (nest,1)
scala> val tupleInt=(1,2,3,4)
tupleInt: (Int, Int, Int, Int) = (1,2,3,4)
```

可以看到，通过(elem1,elem2,....,elemN)的方式可以定义元组。而元组模式匹配的过程与创建过程相反，用于对元组中的内容进行析取，示例如下：

```
/**
 * 元组模式
 */
object ScalaExampel9_9 extends App{
  //定义一个元组
  val tupleInt=(1,2,3,4)
  def patternMatching(x:AnyRef)=x match {
    //元组模式，匹配两个元素的元组
    case (first,second) => println(s"元组中第一个元素=$first,第二个元素=$second")

    //first,three 分别匹配元组中的第一、三个元素
    //第一个_匹配元组中的第二个元素，第二个_匹配元组中的第四个元素
    case (first,_,three,_) => println(s"元组中第一个元素=$first,第三个元素=$three")

    //元组模式中不能使用_*的匹配模式
    //case (first,_*) => println(s"元组中第一个元素=$first")
    case _ =>
  }
  patternMatching(tupleInt)
}
```

代码运行结果如下：

元组中第一个元素=1,第三个元素=3

代码 case (first,_,three,_)会匹配定义的元组 tupleInt，first 会匹配元组中的第一个元素，"_"匹配元组中的第二个元素但不赋值，three 会匹配元组中的第三个元组，"_"会匹配元组中的第四个元素但不赋值，注意元组模式匹配不能使用 case (first,_*)这种形式，即 "_*" 不会匹配元组中的剩余 3 个元素，"_*" 这种匹配方式只能在序列模式中使用，不能在其他模式匹配中使用。

9.2.6 类型模式

类型模式用于匹配变量的类型，Scala 是一种静态类型语言，任何变量在定义时都是有类型的，在不给定的情况下，Scala 会使用类型推导确定变量的类型。模式匹配中的类型模式用于判断变量对应的具体类型，例如：

```
/**
 * 类型模式
 */
object ScalaExampel9_10 extends App{
  //定义一个元组
  class A
  class B extends A
  class C extends A
  val b=new B
  val c=new C
  def patternMatching(x:AnyRef)=x match {
      //类型模式，匹配字符串
    case x:String=>println("字符串类型")
      //类型模式，匹配类B
    case x:B=>println("对象类型为B")
      //类型模式，匹配类A
    case x:A=>println("对象类型为A")
    case _=>println("其他类型")
  }
  patternMatching("Scala Pattern Matching")
  patternMatching(b)
  patternMatching(c)
}
```

代码运行结果：

字符串类型
对象类型为B
对象类型为A

在代码中定义了 3 个类，分别是类 A、B、C，其中 B 和 C 为 A 的子类，类型模式 case x:String 会匹配输入类型是字符串的变量，类型模式 case x:B 会匹配输入类型是 B 类及其子类的对象，case x:A 会匹配输入类型是 A 类及其子类的对象。patternMatching(c)之所以会输出"对象类型为 A"是因为对象 c 的类型为类 A 的子类，它仍然满足 case x:A 这一模式，也就是说类型模式也具有多态性。

9.2.7 变量绑定模式

在前面我们讲构造函数模式时，通过下列代码

```
case Dog(name,age) => println(s"Dog name=$name,age=$age")
```

对输入的对象进行析构，以析取对象的成员域 name 和 age。但在实际应用开发的过程中，如果想要的不是析取对象，而是要返回匹配该模式的对象，此时就需要使用变量绑定模式。例如：

```
/**
 * 变量绑定模式
 */
object ScalaExampel9_11 extends App{
  //定义一个元组
  case class Dog(val name:String,val age:Int)
  val dog=Dog("Pet",2)

  def patternMatching(x:AnyRef)=x match {
    //变量绑定模式，匹配成功，则将整个对象赋值给变量 d
    case d@Dog(_,_)=>println("变量绑定模式返回的变量值为："+d)
    case _ =>
  }
  patternMatching(dog)
}
```

代码运行结果如下：

```
变量绑定模式返回的变量值为：Dog(Pet,2)
```

代码中 case d@Dog(_,_)=>println("变量绑定模式返回的变量值为："+d)为变量绑定模式的用法，Dog(_,_)用于匹配输入对象中包括两个成员变量的 Dog 对象，与构造函数模式不同的是，这里不是要析取对象，而是进行完整的模式匹配，如果匹配成功则将整体对象赋值给变量 d。

变量绑定模式在实际应用开发中十分常见，下面再给出一个实例演示变量绑定模式匹配的应用。

```
/**
 * 变量绑定模式
 */
object ScalaExampel9_12 extends App{
```

```scala
  //定义一个元组
  val list=List(List(1,2,3,4),List(4,5,6,7,8,9))
  def patternMatching(x:AnyRef)=x match {
    //变量绑定模式
    case e1@List(_,e2@List(4,_*))=>println("变量 e1="+e1+"\n,变量 e2="+e2)
    case _ =>
  }
  patternMatching(list)
}
```

代码运行结果：

```
变量 e1=List(List(1, 2, 3, 4), List(4, 5, 6, 7, 8, 9))
变量 e2=List(4, 5, 6, 7, 8, 9)
```

代码 val list=List(List(1,2,3,4),List(4,5,6,7,8,9))定义了一个 List，该 List 有两个子元素且都为 List，变量绑定模式 case e1@List(_,e2@List(4,_*))=>println("变量 e1="+e1+"\n,变量 e2="+e2)中，在变量 e1 返回的是匹配成功的整个 List。List(_,e2@List(4,_*))表示匹配包含两个 List 元素的 List，通配符"_"匹配第一个 List 元素，它可以是任何内容和任何大小的 List。List(4,_*)表示匹配第二个 List 元素，要求该 List 第一个元素为 4 即可。如果匹配成功则将整个 List 赋值给变量 e1，将 List 第二个元素赋值给 e2。

9.3 模式匹配原理

9.3.1 构造函数模式匹配原理

在前面构造函数模式匹配中先定义一个样例类：case class Dog(val name:String,val age:Int)，然后再使用 case Dog(name,age) => println(s"Dog name=$name,age=$age")进行构造函数模式匹配。如果将 Dog 类定义成普通的类：class Dog(val name:String,val age:Int)，则不能使用构造函数模式匹配。例如：

```scala
/**
 * 构造函数模式匹配原理
 */
object ScalaExampel9_13 extends App{

  //定义普通的类 Dog
  class Dog(val name:String,val age:Int)
  //利用构造函数创建对象,这里必须通过显示地通过 new 创建对象
  val dog=new Dog("Pet",2)

  def patternMatching(x:AnyRef)=x match {
    //编译通不过,不能使用构造函数模式
```

```
    case Dog(name,age) => println(s"Dog name=$name,age=$age")
    case _ =>
  }
  patternMatching(dog)
}
```

上面的代码中，case Dog(name,age) => println(s"Dog name=$name,age=$age")处会编译出错：

```
Error:(221, 10) not found: value Dog
    case Dog(name,age) => println(s"Dog name=$name,age=$age")
```

这是因为如果定义一个普通的 Dog 类的话，需要显式地定义 Dog 类的伴生对象并在伴生对象中实现 unapply 方法，具体实现如下：

```
/**
  * 构造函数模式匹配原理：手动在类的伴生对象中实现 unapply 方法
  */
object ScalaExampel9_13 extends App{
  //定义普通的类 Dog
  class Dog(val name:String,val age:Int)
  //Dog 类的伴生对象
  object Dog{
    //手动实现 unapply 方法
    def unapply(dog:Dog):Option[(String,Int)]={
        if(dog!=null) Some(dog.name,dog.age)
        else None
    }
  }
  //利用构造函数创建对象,因为 Dog 是普通的类,这里必须通过显示地 new 来创建对象
  val dog=new Dog("Pet",2)
  def patternMatching(x:AnyRef)=x match {
    //因为在 Dog 的伴生对象中实现了 unapply 方法，此处可以使用构造函数模式
    case Dog(name,age) => println(s"Dog name=$name,age=$age")
    case _ =>
  }
  patternMatching(dog)
}
```

代码运行结果如下：

```
Dog name=Pet,age=2
```

通过上述代码可以看到，除了定义 class Dog(val name:String,val age:Int)类之外，还定义了其伴生对象 Dog 并在该对象中实现了 def unapply(dog:Dog)方法，此时 case Dog(name,age) => println(s"Dog name=$name,age=$age")便能够编译通过且程序能够同 ScalaExampel9_6 一样正确地输出结果。通过上述代码不难看出，构造函数模式是通过 unapply 方法起作用的，unapply 方法最终完成对输入对象的析构。将 Dog 类定义为 case class 之后，便不再需要手动去编写伴

生对象 Dog 并实现其 unapply 方法，这是因为一个类一旦被定义为 case class 后，编译器会自动帮我们生成该类的伴生对象并实现 apply 方法及 unapply 方法。

除构造函数模式外，元组模式匹配也是通过 unapply 方法起作用的，例如两个元素的元组，元组类定义如下：

```
case class Tuple2[@specialized(Int, Long, Double, Char, Boolean/*, AnyRef*/) +T1, @specialized(Int, Long, Double, Char, Boolean/*, AnyRef*/) +T2](_1: T1, _2: T2)
    extends Product2[T1, T2]
```

可以看到，Tuple2 也被定义为 case class，编译器会自动生成对应的 apply 及 unapply 方法，从而可以在元组模式匹配中对元组的内容进行析取。

9.3.2 序列模式匹配原理

在前面的序列模式匹配中，我们特别提到像 case Array(first,_,three,_*)这种模式中的"_*"只能用于序列，不能在元组及其他模式中使用，这是因为序列模式在进行模式匹配时使用的不是 unapply 方法而是 unapplySeq 方法，在 scala.Array 类的伴生对象中，可以看到伴生对象 object Array 的定义如下：

```
object Array extends FallbackArrayBuilding {
  /** Creates an array with given elements.
   *
   *  @param xs the elements to put in the array
   *  @return an array containing all elements from xs.
   */
  // Subject to a compiler optimization in Cleanup.
  // Array(e0, ..., en) is translated to { val a = new Array(3); a(i) = ei; a }
  def apply[T: ClassTag](xs: T*): Array[T] = {
    val array = new Array[T](xs.length)
    var i = 0
    for (x <- xs.iterator) { array(i) = x; i += 1 }
    array
  }

  /** Creates an array of `Boolean` objects */
  // Subject to a compiler optimization in Cleanup, see above.
  def apply(x: Boolean, xs: Boolean*): Array[Boolean] = {
    val array = new Array[Boolean](xs.length + 1)
    array(0) = x
    var i = 1
    for (x <- xs.iterator) { array(i) = x; i += 1 }
    array
```

```
    }
    //其他 apply 方法等省略
    /** Called in a pattern match like `{ case Array(x,y,z) => println('3 elemen
ts')}`.
     *
     * @param x the selector value
     * @return  sequence wrapped in a [[scala.Some]], if `x` is a Seq, otherwi
se `None`
     */
    def unapplySeq[T](x: Array[T]): Option[IndexedSeq[T]] =
      if (x == null) None else Some(x.toIndexedSeq)
```

可以看到 Array 类的伴生对象中有两种重要的方法：apply 方法及 unapplySeq 方法。apply 方法让开发人员可以直接通过 Array(1,2,3,4)这种无 new 的显式方式创建对象，而 unapplySeq 方法则可以让开发人员使用序列模式匹配，在进行像 case Array(first,_,three,_*)这样的序列模式匹配时，便会自动调用 unapplySeq 方法完成对序列的内容析取。

9.4　正则表达式与模式匹配

9.4.1　Scala 正则表达式

所有的现代编程语言都无一例外支持正则表达式，例如 Java、Perl、PHP、Python、JavaScript、Linux Shell 等当下流行的编程语言。Scala 语言与 Java 语言有着良好的互操作性，Scala 语言能够完全使用 Java 语言提供的 API 进行正则表达式处理，例如：

```
/**
 * Scala 调用 Java 语言提供的 API 进行正则表达式处理
 */
object ScalaExampel9_14 extends  App{
    import java.util.regex.Pattern
    //正则表达式待匹配的字符串
    val line = "Hadoop has been the most popular big data " +
            "processing tool since 2005-11-21"
    //正则表达式,用于匹配年-月-日这样的日期格式
    val rgex = "(\\d\\d\\d\\d)-(\\d\\d)-(\\d\\d)"

    // 根据正则表达式创建 Pattern 对象
    val pattern = Pattern.compile(rgex)

    // 创建 Matcher 对象
    val m = pattern.matcher(line);
    if (m.find( )) {
```

```
            //m.group(0)返回整个匹配结果, 即(\d\d\d\d)-(\d\d)-(\d\d)匹配结果
            println(m.group(0))
            //m.group(1)返回第一个分组匹配结果, 即(\d\d\d\d)年匹配结果
            println(m.group(1))
            //m.group(2)返回第二个分组匹配结果, 即(\d\d)月匹配结果
            println(m.group(2))
            //m.group(3)返回第三个分组匹配结果, 即(\d\d)日匹配结果
            println(m.group(3))
        } else {
            println("未找到匹配项")
        }
    }
```

代码运行结果如下:

```
2005-11-21
2005
11
21
```

使用 Java 语言提供的 API 能够满足大部分正则表达式的应用场景, 不过 Scala 语言自身也提供正则表达式处理的 API, 推荐在实际开发中使用, 因为 Scala 语言提供的 API 更灵活、代码更简洁, 更重要的是 Scala 语言提供的正则表达式处理 API 能够与模式匹配完美结合。

在 Scala 语言中有两种创建正则表达式的方式, 一种是通过 r 方法直接将字符串转换成正则表达式对象, 对应类为 scala.util.matching.Regex, 例如:

```
scala> val dateP1 = """(\d\d\d\d)-(\d\d)-(\d\d)""".r
dateP1: scala.util.matching.Regex = (\d\d\d\d)-(\d\d)-(\d\d)
```

通过 r 方法创建 Scala 正则表达式对象, 其作用原理是通过隐式转换将字符串转换为 scala.collection.immutable.StringLike 对象, 然后调用该对象中的 r 方法将字符串转换为 scala.util.matching.Regex 对象。另外一种创建正则表达式对象的方式是直接显式地调用 scala.util.matching.Regex 构造函数创建, 例如:

```
scala> val dateP2 = new scala.util.matching.Regex("""(\d\d\d\d)-(\d\d)-(\d\d)""")
dateP2: scala.util.matching.Regex = (\d\d\d\d)-(\d\d)-(\d\d)
```

代码中直接通过"""(\d\d\d\d)-(\d\d)-(\d\d)"""三引号的方式是为了避免 Java 语言中直接使用转义符"(\\d\\d\\d\\d)-(\\d\\d)-(\\d\\d)"所带来的阅读性差等问题。

使用 Scala 正则表达式, 需要掌握以下几个重要方法:

(1) def findAllIn(source: CharSequence): MatchIterator。该方法用于匹配字符串中所有与输入模式相匹配的字符, 并返回 scala.util.matching.Regex.MatchIterator 对象。MatchIterator 是一种特殊的 scala.collection.Iterator 迭代器, 可以使用 for 循环迭代输出所有匹配的模式, 具体代码如下:

```
scala> val rgex ="""(\d\d\d\d)-(\d\d)-(\d\d)""".r
rgex: scala.util.matching.Regex = (\d\d\d\d)-(\d\d)-(\d\d)

scala> for (date <- rgex findAllIn "2015-12-31 2016-02-20") {
  println(date)
}
2015-12-31
2016-02-20
```

(2) def findAllMatchIn(source: CharSequence): Iterator[Match]。以 Iterator[Match]形式返回所有匹配的模式。

```
scala> val rgex ="""(\d\d\d\d)-(\d\d)-(\d\d)""".r
rgex: scala.util.matching.Regex = (\d\d\d\d)-(\d\d)-(\d\d)

//date 为 Match 对象，groupCount 返回的是匹配模式的分组数
scala> for (date <- rgex findAllMatchIn "2015-12-31 2016-02-20") {
  println(date.groupCount)
}
3
3
```

(3) def findFirstIn(source: CharSequence): Option[String]。返回匹配成功的第一个字符串，匹配成功返回 Option[String]，匹配失败则返回 None。

```
scala> val rgex ="""(\d\d\d\d)-(\d\d)-(\d\d)""".r
rgex: scala.util.matching.Regex = (\d\d\d\d)-(\d\d)-(\d\d)

scala> val copyright: String = rgex findFirstIn "Date of this document: 2011
-07-15" match {
  case Some(date)=> "Copyright "+date
  case None                => "No copyright"
}
copyright: String = Copyright 2011-07-15
```

(4) def findFirstMatchIn(source: CharSequence): Option[Match]。返回匹配成功的每一个字符串，以 Match 对象形式返回，匹配成功返回 Option[Match]，失败则返回 None。该方法与 def findFirstIn(source: CharSequence): Option[String]方法的区别在于它能够获取更多关于匹配的信息，如分组数、匹配字符串的索引等信息。

```
//创建正则表达式对象时，指定模式的分组名称
scala> val dateP2 = new scala.util.matching.Regex("""(\d\d\d\d)-(\d\d)-(\d\d)""", "year", "month", "day")
dateP2: scala.util.matching.Regex = (\d\d\d\d)-(\d\d)-(\d\d)

scala> val result=dateP2 findFirstMatchIn "2015-12-31 2016-02-20" match {
```

```
    //m.group 方法可以返回对应匹配分组的值
    case Some(m)=> "year 对应分组的值为："+m.group("year")
    case None                => "未匹配成功"
}
result:year 对应分组的值为： 2015
```

（5）def replaceAllIn(target: CharSequence, replacer: (Match) ⇒ String): String。使用 replacer 函数对所有模式匹配成功的字符串进行替换，replacer 函数的输入参数为 Match，返回值类型为 String。

```
//创建正则表达式对象，并指定模式的分组名称
scala> val datePattern = new Regex("""(\d\d\d\d)-(\d\d)-(\d\d)""", "year", "month", "day")
datePattern: scala.util.matching.Regex = (\d\d\d\d)-(\d\d)-(\d\d)

//待替换的字符串
scala> val text = "From 2011-07-15 to 2011-07-17"
text: String = From 2011-07-15 to 2011-07-17

//使用 replaceAllIn 函数对匹配成功的字符串进行替换
scala> val repl = datePattern replaceAllIn (text, m => m.group("month")+"/"+m.group("day"))
repl: String = From 07/15 to 07/17
```

def replaceAllIn(target: CharSequence, replacer: (Match) ⇒ String): String 方法还有一个重载的方法 def replaceAllIn(target: CharSequence, replacement: String): String，该方法的作用与 replaceAllIn 作用类似，只不过第二个参数不再是一个函数 replacer: (Match) ⇒ String，而是直接给定的字符串 replacement: String。另外，还有一个重要方法 def replaceFirstIn(target: CharSequence, replacement: String): String，其作用是替换第一个匹配成功的字符串，感兴趣的读者可以查阅官方文档，这里不再赘述。

9.4.2 正则表达式在模式匹配中的应用

在前一节中对 Scala 的正则表达式的用法进行了介绍，并初步演示了 Scala 正则表达式与模式匹配结合使用，例如：

```
val result=dateP2 findFirstMatchIn "2015-12-31 2016-02-20" match {
    //m.group 方法可以返回对应匹配分组的值
    case Some(m)=> "year 对应分组的值为："+m.group("year")
    case None                => "未匹配成功"
}
```

实际上，Scala 正则表达式与模式匹配还有许多更深层次的用法。

（1）提取模式的分组值

在前面的代码中通过 m.group("year")来获取模式的分组值，这种方式不是特别灵活，Scala

正则表达式可以与模式匹配结合起来发挥其强大的功能，下面的代码演示的是如何利用模式匹配直接提取模式中的分组信息。

```scala
/**
  * Scala 正则表达式使用
  * 提取模式的分组值
  */
object ScalaExampel9_16 extends App{
  //正则表达式
  val dateRegx ="""(\d\d\d\d)-(\d\d)-(\d\d)""".r

  //待匹配的字符串
  val text="2015-12-31 2016-02-20"

  //提取模式的分组信息
  for (date<- dateRegx.findAllIn(text)) {
    date match {
      case dateRegx(year,month,day)
=>println(s"match 语句中的模式匹配: year=$year,month=$month,day=$day")
      case _ =>
    }
  }

  //提取模式的分组信息，与前面的代码作用等同
  for (dateRegx(year,month,day)<- dateRegx.findAllIn(text)) {
    println(s"for 循环中的正则表达式模式匹配:year=$year,month=$month,day=$day")
  }
}
```

代码运行结果如下：

```
match 语句中的模式匹配: year=2015,month=12,day=31
match 语句中的模式匹配: year=2016,month=02,day=20
for 循环中的正则表达式模式匹配: year=2015,month=12,day=31
for 循环中的正则表达式模式匹配: year=2016,month=02,day=20
```

通过上面的代码可以看到，首先通过 for (date<- dateRegx.findAllIn(text))得到 MatchIterator 对象并对其进行分组，然后在循环中使用 case dateRegx(year,month,day)将分组信息提取出来，实际上在 for 循环中也可以直接使用模式匹配，即通过代码 for (dateRegx(year,month,day)<- dateRegx.findAllIn(text))进行简化，从而使代码更加简洁，关于 for 循环中的模式匹配我们将在 9.5 小节中进行更深入的介绍。

在使用 findFirstMatchIn 等返回值类型为 Option[Match]的方法时，如果要使用模式匹配提取分组信息，可以使用下列方式：

```scala
/**
  * Scala 正则表达式使用
```

```
 * findFirstMatchIn 等方法与模式匹配
 */
object ScalaExampel9_17 extends App{
  //正则表达式
  val dateRegx ="""(\d\d\d\d)-(\d\d)-(\d\d)""".r

  //待匹配的字符串
  val text="2015-12-31 2016-02-20"

  //findFirstMatchIn 返回值类型为 Option[Match]
  dateRegx.findFirstMatchIn(text) match{
    case Some(dateRegx(year,month,day))=>println(s"findFirstMatchIn 与模式匹配：year=$year,month=$month,day=$day")
    case None=>println("没有找到匹配")
  }

  //findFirstIn 返回值类型为 Option[String]
  dateRegx.findFirstIn(text) match{
    case Some(dateRegx(year,month,day))=>println(s"findFirstIn 与模式匹配：year=$year,month=$month,day=$day")
    case None=>println("没有找到匹配")
  }

}
```

代码运行结果如下：

```
findFirstMatchIn 与模式匹配：year=2015,month=12,day=31
findFirstIn 与模式匹配：year=2015,month=12,day=31
```

dateRegx.findFirstMatchIn(text)返回的对象类型要么是 Option[Match]，要么是 None（无匹配时），如果匹配成功则会使用 case Some(dateRegx(year,month,day))将正则表达式对象的分组信息提取出来。而 dateRegx.findFirstIn(text)返回的对象要么是 Option[String]，要么是 None，如果匹配成功则会使用 case Some(dateRegx(year,month,day))将正则表达式对象的分组信息提取出来。

通过前面的代码可以看到，dateRegx.findFirstIn(text)和 dateRegx.findFirstMatchIn(text)虽然返回的类型不一样，但都可以使用 dateRegx(year,month,day)这种模式匹配方式对分组信息进行提取，这是为什么呢？正则表达式中的模式匹配其背后的作用原理又是什么呢？事实上，正则表达式的模式匹配与序列模式的模式匹配作用原理类似，正则表达式也是通过 unapplySeq 方法起作用的，将正则表达式对象进行析构，完成分组信息的提取，在 scala.util.util.Regex 对象中可以查看 unapplySeq 方法的定义：

```
/** Tries to match target (whole match) and returns the matching subgroups.
 *  if the pattern has no subgroups, then it returns an empty list on a
 *  successful match.
```

```
 *
 * Note, however, that if some subgroup has not been matched, a `null` will
 * be returned for that subgroup.
   其他注释省略
 */
def unapplySeq(target: Any): Option[List[String]] = target match {
  case s: CharSequence =>
    val m = pattern matcher s
    if (runMatcher(m)) Some((1 to m.groupCount).toList map m.group)
    else None
  case m: Match => unapplySeq(m.matched)
  case _ => None
}
```

从 unapplySeq 方法的定义来看，它可以处理输入类型为 CharSequence 和 Match 的正则表达式匹配结果，这就是为什么它能够对 dateRegx.findFirstIn(text) 和 dateRegx.findFirstMatchIn(text)返回的不同类型进行处理而代码处理逻辑相同的原因。

9.5 for 循环中的模式匹配

在 9.4 节中提到，在 for 循环中可以直接使用模式匹配，如 for (dateRegx(year,month,day)<-dateRegx.findAllIn(text))可以直接完成正则表达式模式匹配。在本节当中，我们将从模式匹配的视角来理解 for 循环并进一步体会模式匹配的强大。

下面的代码是普通 for 循环的使用：

```
scala>
//使用 for 循环，给变量进行赋值
for(i<- 1 to 5){
  print(i+" ")
}
println

scala>  1 2 3 4 5
```

接下来将通过模式匹配的视角对 for 循环的使用进行更深入的理解。

（1）变量模式匹配

下面的代码给出的是在 for 循环中使用变量模式匹配给变量赋值的使用方法：

```
//使用 for 循环，模式匹配的视角（变量模式匹配）
for((language,framework)<-Map("Java"->"Hadoop","Closure"->"Storm","Scala"->"Spark")){
    println(s"$framework is developed by $language language")
```

```
            }
```

代码运行结果如下:

```
Hadoop is developed by Java language
Storm is developed by Closure language
Spark is developed by Scala language
```

(2) 常量模式匹配

下面的代码给出的是 for 循环中常量模式匹配的使用方法。

```
//使用 for 循环，模式匹配的视角，(常量模式匹配)
    for((language,"Spark")<-Map("Java"->"Hadoop","Closure"->"Storm","Scala"->"Spark")){
        println(s"Spark is developed by $language language")
    }
```

代码运行结果如下:

```
Spark is developed by Scala language
```

(3) 变量绑定模式匹配

下面的代码给出的是变量绑定模式匹配的使用方法。

```
//使用 for 循环，模式匹配的视角，(变量绑定模式匹配)
    for((language,e@"Spark")<-Map("Java"->"Hadoop","Closure"->"Storm","Scala"->"Spark")){
        println(s"$e is developed by $language language")
    }
```

代码运行结果如下:

```
Spark is developed by Scala language
```

(4) 类型模式匹配

下面的代码给出的是类型模式匹配的使用方法。

```
//使用 for 循环，模式匹配的视角，(类型模式匹配)
    for((language,framework:String)<-Map("Java"->"Hadoop".length,"Closure"->"Storm".length,"Scala"->"Spark")){
        println(s"$language is developed by $language language")
    }
```

代码运行结果如下:

```
Spark is developed by Scala language
```

(5) 构造函数模式匹配

下面的代码给出的是 for 循环中构造函数模式匹配的使用方法。

```
//使用 for 循环，模式匹配的视角，(构造函数模式匹配)
    case class Dog(val name:String,val age:Int)
```

```
  for(Dog(name,age)<-List(Dog("Pet",2),Dog("Penny",3),Dog("Digo",2))){
    println(s"Dog $name is  $age years old")
  }
```

代码运行结果如下：

```
Dog Pet is  2 years old
Dog Penny is  3 years old
Dog Digo is  2 years old
```

（6）序列模式匹配

下面的代码给出的是 for 循环中序列模式匹配的使用方法。

```
//使用 for 循环，模式匹配的视角，(序列模式匹配)
  for(List(first,_*)<-List(List(1,2,3),List(4,5,6,7))){
     println(s"the first elemement is $first")
  }
```

代码运行结果如下：

```
the first elemement is 1
the first elemement is 4
```

9.6 模式匹配与样例类、样例对象

9.6.1 模式匹配与样例类

在 9.3 节中曾提到，如果只定义一个普通类如 Dog 类，则需要手动定义其 Dog 类的伴生对象并在伴生对象中实现相应的 unapply 方法，只有这样才能使用 Dog 类的构造函数模式匹配。但如果将 Dog 类定义为样例类后，编译器会自动帮我们创建其伴生对象并实现相应的 unapply 方法，从而避免大量手动编写代码所带来的程序复杂性。事实上，模式匹配与样例类是一对孪生兄弟，在模式匹配中经常一起使用。

在 Scala 的实际开发过程中，我们经常会遇到这种情况：假设类或特质 A 有且仅有 4 个子类，分类为 B1、B2、B3、B4，在进行模式匹配时，常常需要列出所有 4 种子类的情况，从而避免由没有被处理的情况而带来的程序出错。此时，A 最好被声明为 sealed trait A 或 sealed class A，当使用模式匹配未列出所有子类匹配情况时，编译器会出给警告。为说明这种使用方式，下面举个具体的例子。

```
/**
  * 模式匹配与样例类
  */
object ScalaExampel9_19 extends  App{
  //定义一个 sealed trait DeployMessage
  sealed trait DeployMessage
```

```scala
    //定义三个具体的子类,全部为 case class
    case class RegisterWorker(id: String, host: String, port: Int) extends DeployMessage
    case class UnRegisterWorker(id: String, host: String, port: Int) extends DeployMessage
    case class Heartbeat(workerId: String) extends DeployMessage

    //handleMessage 函数会处理所有可能的情况,即穷举出所有 DeployMessage 的子类
    def handleMessage(msg:DeployMessage)= msg match {
      case RegisterWorker(id,host,port)=>s"The worker $id is registering on $host:$port"
      case UnRegisterWorker(id,host,port)=>s"The worker $id is unregistering on $host:$port"
      case Heartbeat(id)=>s"The worker $id is sending heartbeat"
    }

    val msgRegister=RegisterWorker("204799","192.168.1.109",8079)
    // val msgUnregister=UnRegisterWorker("204799","192.168.1.109",8079)
    // val msgHeartbeat=Heartbeat("204799")

    println(handleMessage(msgRegister))

  }
```

代码运行结果如下:

```
The worker 204799 is registering on 192.168.1.109:8079
```

在上面的代码中,DeployMessage 被声明为 sealed trait,然后为其定义了 3 个具体的子类,分别是 case class RegisterWorker(id: String, host: String, port: Int)、case class UnRegisterWorker(id: String, host: String, port: Int)、case class Heartbeat(workerId: String)。def handleMessage(msg:DeployMessage)利用模式匹配处理 DeployMessage 对应所有种类的消息,如果函数中任意一匹配项注释掉的话,编译器就会发出警告。例如将代码 case Heartbeat(id)=>s"The worker $id is sending heartbeat"注释掉,注释后的代码如下:

```scala
    def handleMessage(msg:DeployMessage)= msg match {
      case RegisterWorker(id,host,port)=>s"The worker $id is registering on $host:$port"
      case UnRegisterWorker(id,host,port)=>s"The worker $id is unregistering on $host:$port"
      //case Heartbeat(id)=>s"The worker $id is sending heartbeat"
    }
```

此时,对 ScalaExampel9_19 重新进行编译时,编译器会给出下列警告:

```
warning: match may not be exhaustive.It would fail on the following input: H
eartbeat(_)
```

这是定义 sealed trait 或 sealed class 所带来的好处,推荐在子类众多,模式匹配处理较复杂且需要处理所有可能发生的情况时使用。

9.6.2 模式匹配与样例对象

在代码 ScalaExampel9_19 中,我们定义了 DeployMessage 及其 3 个子类 RegisterWorker、UnRegisterWorker 和 Heartbeat,这 3 个子类都有自己的成员域,但在实际程序开发时,DeployMessage 的部分子类可能并不需要定义自己的成员域,而只是用于消息的标识。假设该消息为 RequestWorkerState,此时如果仍然将其定义为 case class 的话,编译器会给出警告,例如:

```
// RequestWorkerState 没有成员域,编译器会给出警告:
// case class without a parameter list is deprecated
case class RequestWorkerState extends DeployMessage
```

此时推荐使用样例对象(case object)来声明 RequestWorkerState,完整代码如下:

```
/**
 * 模式匹配与样例对象
 */
object ScalaExampel9_20 extends App{
  //定义一个 sealed trait DeployMessage
  sealed trait DeployMessage

  //定义3个具体的子类,全部为 case class
  case class RegisterWorker(id: String, host: String, port: Int) extends DeployMessage
  case class UnRegisterWorker(id: String, host: String, port: Int) extends DeployMessage
  case class Heartbeat(workerId: String) extends DeployMessage
  case object RequestWorkerState extends DeployMessage
  //handleMessage 函数会处理所有可能的情况,即穷举出所有 DeployMessage 的子类
  def handleMessage(msg:DeployMessage)= msg match {
    case RegisterWorker(id,host,port)=>s"The worker $id is registering on $host:$port"
    case UnRegisterWorker(id,host,port)=>s"The worker $id is unregistering on $host:$port"
    case Heartbeat(id)=>s"The worker $id is sending heartbeat"
    case RequestWorkerState=>"Request Worker State"
  }

  val msgRegister=RegisterWorker("204799","192.168.1.109",8079)
```

```
    //  val msgUnregister=UnRegisterWorker("204799","192.168.1.109",8079)
    //  val msgHeartbeat=Heartbeat("204799")

    println(handleMessage(msgRegister))
    println(handleMessage(RequestWorkerState))

}
```

代码运行结果如下：

```
The worker 204799 is registering on 192.168.1.109:8079
Request Worker State
```

通过上述代码可以看到，将 RequestWorkerState 声明为 case object 之后，可以直接将其传递给函数 handleMessage，而不需要像 RegisterWorker 等 case class 类一样，必须先创建相应的对象，这是将 RequestWorkerState 定义为 case object 的好处。为什么要使用 case object 而不推荐使用 case class 呢？让我们来深入分析 case class 与 case object 的差异，先来看一下 object ScalaExampel9_20 编译后生成的字节码文件：

```
    E:\IntellijIDEAWorkspace\out\production\ScalaProject\cn\scala\chapter09 的
目录
    016/02/21  00:16            4,542 ScalaExampel9_20$.class
    016/02/21  00:16            1,533 ScalaExampel9_20$delayedInit$body.class
    016/02/21  00:16              243 ScalaExampel9_20$DeployMessage.class
    016/02/21  00:16            1,813 ScalaExampel9_20$Heartbeat$.class
    016/02/21  00:16            2,549 ScalaExampel9_20$Heartbeat.class
    016/02/21  00:16            2,358 ScalaExampel9_20$RegisterWorker$.class
    016/02/21  00:16            3,253 ScalaExampel9_20$RegisterWorker.class
    016/02/21  00:16            1,944 ScalaExampel9_20$RequestWorkerState$.clas
s
    016/02/21  00:16            2,378 ScalaExampel9_20$UnRegisterWorker$.class
    016/02/21  00:16            3,261 ScalaExampel9_20$UnRegisterWorker.class
    016/02/21  00:16            4,749 ScalaExampel9_20.class
```

上述生成的字节码文件目录中，ScalaExampel9_20$DeployMessage.class 对应我们的 sealed trait DeployMessage，对其进行反编译得到如下代码：

```
    E:\IntellijIDEAWorkspace\out\production\ScalaProject\cn\scala\chapter09>ja
vap -private ScalaExampel9_20$DeployMessage.class
    Compiled from "chapter09.scala"
    public interface cn.scala.chapter09.ScalaExampel9_20$DeployMessage {
    }
```

case class Heartbeat(workerId: String) extends DeployMessage 对应的字节码文件为 ScalaExampel9_20$Heartbeat$.class、ScalaExampel9_20$Heartbeat.class，对它们进行反编译得到如下代码：

```
E:\IntellijIDEAWorkspace\out\production\ScalaProject\cn\scala\chapter09>ja
vap -private ScalaExampel9_20$Heartbeat$.class
Compiled from "chapter09.scala"
public class cn.scala.chapter09.ScalaExampel9_20$Heartbeat$ extends scala.
runtim
e.AbstractFunction1<java.lang.String, cn.scala.chapter09.ScalaExampel9_20$
Heartb
eat> implements scala.Serializable {
  public static final cn.scala.chapter09.ScalaExampel9_20$Heartbeat$ MODULE
$;
  public static {};
  public final java.lang.String toString();
  public cn.scala.chapter09.ScalaExampel9_20$Heartbeat apply(java.lang.Str
ing);
  public scala.Option<java.lang.String> unapply(cn.scala.chapter09.ScalaEx
ampel9
_20$Heartbeat);
  private java.lang.Object readResolve();
  public java.lang.Object apply(java.lang.Object);
  public cn.scala.chapter09.ScalaExampel9_20$Heartbeat$();
}

E:\IntellijIDEAWorkspace\out\production\ScalaProject\cn\scala\chapter09>ja
vap -private ScalaExampel9_20$Heartbeat.class
Compiled from "chapter09.scala"
public class cn.scala.chapter09.ScalaExampel9_20$Heartbeat implements cn.s
cala.c
hapter09.ScalaExampel9_20$DeployMessage,scala.Product,scala.Serializable {
  private final java.lang.String workerId;
  public java.lang.String workerId();
  public cn.scala.chapter09.ScalaExampel9_20$Heartbeat copy(java.lang.Stri
ng);
  public java.lang.String copy$default$1();
  public java.lang.String productPrefix();
  public int productArity();
  public java.lang.Object productElement(int);
  public scala.collection.Iterator<java.lang.Object> productIterator();
  public boolean canEqual(java.lang.Object);
  public int hashCode();
  public java.lang.String toString();
  public boolean equals(java.lang.Object);
  public cn.scala.chapter09.ScalaExampel9_20$Heartbeat(java.lang.String);
}
```

从反编译后得到的结果可以看到，ScalaExampel9_20$Heartbeat$.class 对应的是 Heartbeat 伴生对象所生成的字节码文件，它自动地生成了 apply 和 unapply 等方法，ScalaExampel9_20$Heartbeat.class 对应的是 Heartbeat 伴生类所生成的字节码文件。而 case object RequestWorkerStateScala 最终只生成了 Exampel9_20$RequestWorkerState$.class 一个字节码文件，其反编译后的内容如下：

```
E:\IntellijIDEAWorkspace\out\production\ScalaProject\cn\scala\chapter09>ja
vap -private ScalaExampel9_20$RequestWorkerState$.class
    Compiled from "chapter09.scala"
    public class cn.scala.chapter09.ScalaExampel9_20$RequestWorkerState$ imple
ments
    cn.scala.chapter09.ScalaExampel9_20$DeployMessage,scala.Product,scala.Seri
alizab
le {
    public static final cn.scala.chapter09.ScalaExampel9_20$RequestWorkerStat
e$ MODULE$;
    public static {};
    public java.lang.String productPrefix();
    public int productArity();
    public java.lang.Object productElement(int);
    public scala.collection.Iterator<java.lang.Object> productIterator();
    public boolean canEqual(java.lang.Object);
    public int hashCode();
    public java.lang.String toString();
    private java.lang.Object readResolve();
    public cn.scala.chapter09.ScalaExampel9_20$RequestWorkerState$();
    }
```

对比 ScalaExampel9_20$Heartbeat$.class 对应的反编译后的代码可以看到，ScalaExampel9_20$RequestWorkerState$.class 中没有 apply 和 unapply 方法。

通过上述内容，我们可以对 case class 与 case object 的差异进行总结：

（1）在模式匹配应用时，case class 需要先创建对象，而 case object 可以直接使用。
（2）case class 类会生成两个字节码文件，而 case object 中会生成一个字节码文件。
（3）case class 生成的伴生对象会自动实现 apply 及 unapply 方法，而 case object 中不会。

可以看到，使用 case object 可以提升程序的执行速度，减少编译器的额外开销。

小 结

本章详细介绍了 Scala 语言中的模式匹配，包括模式匹配的 7 种常用类型如常量模式、变量模式、构造函数模式、类型模式、序列模式、元组模式及变量绑定模式，深入剖析了模式匹配背后深层次的原理。详细介绍了正则表达式与模式匹配的使用，从模式的角度理解 for 循环控制结构，对样例类、样例对象在模式匹配中的应用进行了详细介绍。在下一章中，将详细介绍 Scala 语言的隐式转换。

第 10 章　隐式转换

　　Scala 中的隐式转换无处不在，当类型不匹配时常常会自动发生隐式转换。本章将详细介绍隐式转换函数、隐式类与隐式对象、隐式参数与隐式值及隐式转换规则。

10.1　隐式转换简介

隐式转换是 Scala 语言提供的一种强大的语法特性,是学习 Scala 必须要掌握的技能。与 Scala 中的模式匹配一样,Scala 的隐式转换也是无处不在的。在实际开发过程中常常会自动地使用隐式转换,例如在 for 循环中经常使用 for(i<- 1 to 5)这种方式构造循环语句,我们知道 1 to 5 其实是方法调用即 1.to(5),但 Int 类型并不存在 to 这个方法,在执行时会自动进行隐式转换,将 Int 类型转换为 scala.runtime.RichInt。

查看 RichInt 的 API,在 Type Hierarchy 中可以看到如图 10-1 所示的继承关系图,Int 对象通过隐式转换成 RichInt 对象后,再调用 RichInt 对象的 def to(end: Int): Range.Inclusive = Range.inclusive(self, end)方法创建一个 scala.collection.immutable.Range.Inclusive 集合对象,代码如下:

```
scala> 1 to 5
res0: scala.collection.immutable.Range.Inclusive = Range(1, 2, 3, 4, 5)
```

然后通过 for 循环对集合进行遍历,达到程序执行目的。

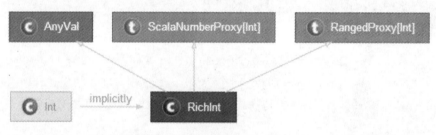

图 10-1　查看 RichInt 的 API

不但如此,Int 类型还可以由其他类型(如 Byte、Char、Short)隐式转换而来,也可以隐式转换为其他类(AnyRef、Integer、Double、Float 及 Long),如图 10-2 所示。

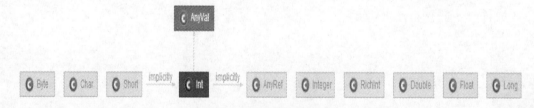

图 10-2　Int 类型隐式转换关系

在第 8 章中介绍 Scala 包的时候曾提到,Scala 会默认引入 Scala 包及 Predef 对象,scala.Int 的伴生对象中包含了 Int 到 Double、Float、Long 的隐式转换函数,如下所示:

```
implicit def int2long(x: Int): Long = x.toLong
  implicit def int2float(x: Int): Float = x.toFloat
  implicit def int2double(x: Int): Double = x.toDouble
```

Int 类型到 RichInt 类型的转换定义在 scala.LowPriorityImplicits 类中。

```
@inline implicit def intWrapper(x: Int)         = new runtime.RichInt(x)
```

其他如 Integer、AnyVal 的隐式转换函数则定义在 Predef 对象中。除此之外，Predef 对象中还定义了大量的其他隐式转换函数，如 Scala 数值类型到 Java 数值封装类型的转换，部分代码如下：

```
implicit def byte2Byte(x: Byte)          = java.lang.Byte.valueOf(x)
 implicit def short2Short(x: Short)       = java.lang.Short.valueOf(x)
 implicit def char2Character(x: Char)     = java.lang.Character.valueOf(x)
 implicit def int2Integer(x: Int)         = java.lang.Integer.valueOf(x)
 implicit def long2Long(x: Long)          = java.lang.Long.valueOf(x)
 implicit def float2Float(x: Float)       = java.lang.Float.valueOf(x)
 implicit def double2Double(x: Double)    = java.lang.Double.valueOf(x)
 implicit def boolean2Boolean(x: Boolean) = java.lang.Boolean.valueOf(x)
```

正是这些隐式转换函数的存在，简化了 Scala 程序代码，使我们编写的代码更简洁。

10.2 隐式转换函数

10.2.1 隐式转换函数的定义

在前一小节，我们提到 Scala 默认已经帮我们实现了大量的隐式转换函数，在特定情况下会自动调用相应的隐式转换函数完成隐式转换，从而保证程序的顺利执行。但 Scala 提供的隐式转换函数数量毕竟是有限的，不可能满足实际应用的所有场景，此时需要定义自己的隐式转换函数。

我们知道，Scala 已经帮我们实现了 Int 类型到 Float 类型的隐式转换函数，例如：

```
implicit def int2float(x: Int): Float = x.toFloat
```

可以直接将整型值赋值给 Float 类型变量：

```
scala> var x:Float=1
x: Float = 1.0
```

现在如果我们想要完成 Float 类型到 Int 类型转换的话，则需要定义相应的隐式转换函数，具体代码如下：

```
//直接将 Float 对象赋值给 Int 类型变量会报错
scala> val intValue:Int=2.55f
<console>:7: error: type mismatch;
 found   : Float(2.55)
 required: Int
    val intValue:Int=2.55f
                     ^
```

```
//定义一个隐式转换函数，该隐式转换函数将Float类型转换成Int类型（注意函数的输入参数）
scala> implicit def float2int(x:Float)=x.toInt
float2int: (x: Float)Int

//再次将Float类型对象赋值给Int类型变量时，程序运行通过
scala> val intValue:Int=2.55f
intValue: Int = 2
```

定义完隐式转换函数 implicit def float2int(x:Float)=x.toInt 后，val intValue:Int=2.55f 语句便能够顺利通过，其原因是：编译器发现赋值对象的类型与最终类型不匹配时，会在当前作用域范围内查找能够将 Float 类型转换成 Int 类型的隐式转换函数，隐式转换函数 float2int 正好满足要求，从而程序可以正常运行。而未定义该隐式转换函数之前，val intValue:Int=2.55f 会出错，这是因为在当前作用域中找不到 Float 类型到 Int 类型的隐式转换函数。

10.2.2 隐式转换函数名称

隐式转换函数 implicit def float2int(x:Float)=x.toInt 中的函数名可以是任意的，即隐式转换函数与函数名称无关，只与函数签名（输入参数类型与返回值类型）有关，如果在当前作用域范围存在函数签名相同但函数名称不同的两个隐式转换函数，则在进行隐式转换时会报错。例如：

```
//定义一个名称为 float2int 的隐式转换函数，输入参数类型为 Float，返回值类型为 Int
scala> implicit def float2int(x:Float)=x.toInt
float2int: (x: Float)Int

//定义一个名称为 f2i 的隐式转换函数，输入参数类型为 Float，返回值类型为 Int
scala> implicit def f2i(x:Float)=x.toInt
f2i: (x: Float)Int

//执行隐式转换时会报错，提示有歧义，因为两个隐式转换函数都能实现 Float 到 Int 的转换
scala> val intValue:Int=2.55f
<console>:10: error: type mismatch;
 found   : Float(2.55)
 required: Int
Note that implicit conversions are not applicable because they are ambiguous:
 both method float2int in object $iw of type (x: Float)Int
 and method f2i in object $iw of type (x: Float)Int
 are possible conversion functions from Float(2.55) to Int
       val intValue:Int=2.55f
                        ^
```

将命令行关闭，然后重新启动，执行下列代码：

```
//当前作用域中只定义了一个从 Float 到 Int 的隐式转换函数
```

```
scala> implicit def f2i(x:Float)=x.toInt
f2i: (x: Float)Int

//成功完成 Float 到 Int 的赋值
scala> val intValue:Int=2.55f
intValue: Int = 2
```

不难看出,隐式转换函数确实只与函数签名有关,而与函数名称无关。虽然如此,但为便于代码的理解和维护,在实际应用中建议给隐式转换函数取能够说明其作用的函数名称。

10.3 隐式类与隐式对象

10.3.1 隐式类

在 10.2 节中,讲解了隐式转换函数是如何将一个类型转换成另外一个类型的,本节要讲的隐式类,其隐式转换却不像隐式转换函数那样明显。

下面通过一个实例说明隐式类的定义与使用,具体代码如下:

```
//通过 implicit 关键字定义一个隐式类,该隐式类的主构造函数接收一个 String 类型的参数
//可以完成 String 类型到 Dog 类型的隐式转换
scala> implicit class Dog(val name:String){
  def bark=println(s"$name is barking" )
}
defined class Dog

//String 类型并不存在 bark 方法,但由于定义的隐式类会进行隐式转换,因此编译器会
//将 String 类型转换成 Dog 类型,再执行 bark 方法
scala> "Nacy".bark
Nacy is barking
```

代码中通过 implicit class Dog(val name:String)定义一个隐式类,可以看到与一般普通的类 class Dog(val name:String)所不同的只是在类前面加了个 implicit 关键字,主构造函数中有一个参数类型为 String 类型的输入参数。当执行代码"Nacy".bark 时,由于 String 类型中并不存在 bark 这个方法,因此编译器会尝试隐式转换。由于 Dog 类是隐式类且主构造函数参数为 String 类型,此时便会调用隐式类 Dog 生成的隐式转换方法完成到 Dog 类的转换,然后再调用其 bark 方法。

那隐式类是如何起作用的?为什么说代码"Nacy".bark 执行的时候,会调用隐式类 Dog 生成的隐式转换方法?事实上隐式类 Dog

```
implicit class Dog(val name:String){
  def bark=println(s"$name is barking" )
}
```

最终会被翻译成下面的代码：

```
class Dog(val name:String){
  def bark=println(s"$name is barking" )
}
implicit def string2Dog(name:String):Dog=new Dog(name)
```

可以看到，隐式类的作用原理是通过隐式转换函数来实现的，执行"Nacy".bark 代码时会调用生成的隐式转换函数 string2Dog 来完成 String 类型到 Dog 类型的转换。不难发现，隐式类所带来的好处是代码更简洁，将类和方法绑定在一起，这样在项目中使用时只需要引入该隐式类即可，不过隐式类带来代码简洁的同时也会使代码更"迷幻"，相比于显式定义隐式转换函数，隐式类中的隐式转换隐藏得更深。

需要注意的是，隐式类的主构造函数参数有且仅有一个，例如：

```
scala> implicit class Dog(val name:String,val age:Int)
<console>:9: error: implicit classes must accept exactly one primary constructor parameter
       implicit class Dog(val name:String,val age:Int)
                      ^

scala> implicit class Dog
<console>:9: error: implicit classes must accept exactly one primary constructor parameter'
       implicit class Dog
                      ^
```

之所以只能有一个参数，是因为隐式转换是将一种类型转换成另外一种类型，源类型与目标类型是一一对应的。

10.3.2 隐式对象

Scala 中除隐式类外，还存在隐式对象，其定义同一般的单例对象定义类似，只不过需要在对象前面再加个 implicit 关键字，示例如下：

```
/*
*隐式对象
*/
object Example10_1 extends  App{
  //定义一个 trait Multiplicable
  trait Multiplicable[T] {
    def multiply(x: T): T
  }
  //定义一个隐式对象 MultiplicableInt,用于整型数据的相乘
    implicit object MultiplicableInt extends Multiplicable[Int] {
      def multiply(x: Int) = x*x
```

```
    }
    //定义一个隐式对象MultiplicableString，用于字符串数据的乘积
    implicit object MultiplicableString extends Multiplicable[String] {
        def multiply(x: String) = x*2
    }
    //定义一个函数，函数具有泛型参数
    def multiply[T: Multiplicable](x:T) = {
        //implicitly 方法，访问隐式对象
        val ev = implicitly[Multiplicable[T]]
        //根据具体的类型调用相应的隐式对象中的方法
        ev.multiply(x)
    }
    //调用隐式对象MultiplicableInt 中的multiply方法
    println(multiply(5))
    //调用隐式对象MultiplicableString 中的multiply方法
    println(multiply("5"))
}
```

代码运行结果如下：

```
25
55
```

代码中 implicit object MultiplicableInt extends Multiplicable[Int]定义了隐式对象，def multiply(x: Int) = x*x 方法给出了其方法的实现，隐式对象 MultiplicableString 的定义与隐式对象 MultiplicableInt 一样，def multiply[T: Multiplicable](x:T)成员方法使用了泛型参数 T: Multiplicable，该泛型参数为上下文界定（泛型与上下文界定将在第 11 章中详细介绍），指在当前作用域必须存在一个类型为 Multiplicable[T]的隐式对象或隐式值（关于隐式值将在 10.4 节在介绍），在 multiply 成员方法中，代码 val ev = implicitly[Multiplicable[T]]调用的是 implicitly 函数（关于 implicitly 函数的作用将在 10.4 节中介绍），该方法将返回最终的隐式对象 ev，代码 ev.multiply(x)会调用具体的方法完成计算。

10.4 隐式参数与隐式值

10.4.1 隐式参数

在 10.3.2 小节中的 Example10_1 示例代码中，我们在 val ev = implicitly[Multiplicable[T]]代码中调用 implicitly 函数来确定最终的隐式对象，该函数也被定义在 Predef 对象中，示例如下：

```
@inline def implicitly[T](implicit e: T) = e
```

可以看到，implicitly 函数中有一个参数 implicit e: T，与一般函数所不同的是，参数 e 前

面也使用了 implicit 关键字修饰，我们称这种形式的参数为隐式参数。通过 val ev = implicitly[Multiplicable[T]]可以看到，调用的时候并没有指定该隐式参数，那值是怎么传进来的呢？这便是隐式参数的作用，前面提到函数 def multiply[T: Multiplicable](x:T)要求在当前作用域存在一个类型为 Multiplicable[T]的隐式值或隐式对象，在调用该函数时，具体的参数类型被确定，如果调用 multiply(5)时参数类型为 Int，则会在当前作用域内查找类型为 Multiplicable[T]的隐式值或隐式对象，此时 MultiplicableInt 满足要求，因此最终 implicitly 方法返回的对象为 MultiplicableInt，然后调用 def multiply(x: Int) = x*x 方法得到最终结果。其实 Example10_1 中的 multiply 方法还可以利用隐式参数进行进一步简化，也就是把以下代码

```scala
//定义一个函数，函数具有泛型参数
  def multiply[T: Multiplicable](x:T) = {
    //implicitly方法，访问隐式对象
    val ev = implicitly[Multiplicable[T]]
    //根据具体的类型调用相应的隐式对象中的方法
    ev.multiply(x)
  }
```

修改为：

```scala
//使用隐式参数定义multiply函数
  def multiply[T: Multiplicable](x:T)(implicit  ev:Multiplicable[T]) = {
    //根据具体的类型调用相应的隐式对象中的方法
    ev.multiply(x)
  }
```

完整代码如下：

```scala
/*
*隐式参数
*/
object Example10_2 extends  App{
  //定义一个trait Multiplicable
  trait Multiplicable[T] {
    def multiply(x: T): T
  }
  //定义一个隐式对象MultiplicableInt，用于整型数据的相乘
  implicit object MultiplicableInt extends Multiplicable[Int] {
    def multiply(x: Int) = x*x
  }
  //定义一个隐式对象MultiplicableString，用于字符串数据的乘积
  implicit object MultiplicableString extends Multiplicable[String] {
    def multiply(x: String) = x*2
  }
  //使用隐式参数定义multiply函数
  def multiply[T: Multiplicable](x:T)(implicit  ev:Multiplicable[T]) = {
    //根据具体的类型调用相应的隐式对象中的方法
```

```
    ev.multiply(x)
  }
  //调用隐式对象 MultiplicableInt 中的 multiply 方法
  println(multiply(5))
  //调用隐式对象 MultiplicableString 中的 multiply 方法
  println(multiply("5"))
}
```

代码运行结果与 Example10_1 相同，函数 def multiply[T: Multiplicable](x:T)(implicit ev:Multiplicable[T])使用隐式参数，它会在当前作用域内查找类型为 Multiplicable[T]的隐式对象，因此在执行 multiply(5)时，由于泛型参数类型为 Int，它便会在当前作用域内查找 Multiplicable[Int]类型的隐式对象，只有 implicit object MultiplicableInt extends Multiplicable[Int]满足要求，此时将 MultiplicableInt 赋值给 ev，然后再调用 def multiply(x: Int) = x*x 执行最终的计算。

10.4.2 隐式值

前面提到，调用 def multiply[T: Multiplicable](x:T)(implicit ev:Multiplicable[T])和 def implicitly[T](implicit e: T) = e 方法时，会在当前作用域内查找相应类型的隐式值或隐式对象，隐式对象在前面已经介绍过，现在让我们来看看什么是隐式值，隐式值也可以称为隐式变量，与一般变量定义的区别是它需要使用 implicit 关键字修饰，例如：

```
scala> implicit val x:Double=2.55
x: Double = 2.55
```

隐式值的作用与隐式对象的作用类似，当函数中有隐式参数时可以不指定参数而将该隐式值传入，例如：

```
scala> implicit val x:Double=2.55
x: Double = 2.55
//定义一个函数 sqrt，该函数的参数为隐式参数
scala> def sqrt(implicit x:Double)=Math.sqrt(x)
sqrt: (implicit x: Double)Double
//调用 sqrt 时不指定参数，此时编译器会在当前作用域内查找相应的隐式值或隐式对象
//这里查找到的是隐式值 x，因此隐式值 x 被当作参数传递给 sqrt，相当于调用 sqrt(x)
scala> sqrt
res0: Double = 1.5968719422671311
```

代码中 def sqrt(implicit x:Double)=Math.sqrt(x)定义了一个函数 sqrt，implicit x:Double 指定其参数为隐式参数，在执行 sqrt 时未指定其参数，此时编译器便会在当前作用域内查找与参数类型匹配的隐式对象或隐式值，显然这里与参数类型匹配的为隐式值 x，因此 x 被当作参数传递即调用 sqrt 相当于调用 sqrt(x)。

熟悉了隐式值的作用后，便可以通过隐式值对 Example10_2 进行改造，请看下列示例：

```
/*
```

```scala
/*使用隐式值对Example10_2进行改造
*/
object Example10_3 extends  App{
  //定义一个trait Multiplicable
  trait Multiplicable[T] {
    def multiply(x: T): T
  }
  //定义一个普通类MultiplicableInt,用于整型数据的相乘
  class MultiplicableInt extends Multiplicable[Int] {
    def multiply(x: Int) = x*x
  }
  //定义一个普通类MultiplicableString,用于字符串数据的乘积
  class MultiplicableString extends Multiplicable[String] {
    def multiply(x: String) = x*2
  }
  //使用隐式参数定义multiply函数
  def multiply[T: Multiplicable](x:T)(implicit  ev:Multiplicable[T]) = {
    //根据具体的类型调用相应的隐式对象中的方法
    ev.multiply(x)
  }

  //类型为MultiplicableInt 的隐式值mInt
  implicit val mInt=new MultiplicableInt

  //类型为MultiplicableString 的隐式值mStr
  implicit val mStr=new MultiplicableString

  //隐式值mInt当作参数传入ev,相当于调用multiply(5)(mInt)
  println(multiply(5))
  //隐式值mStr当作参数传入ev,相当于调用multiply(5)(mStr)
  println(multiply("5"))

}
```

代码 Example10_3 的运行结果与 Example10_2 运行结果相同，与代码 Example10_2 所不同的是，MultiplicableInt 类和 MultiplicableString 类都被定义为普通的类而不是隐式对象，然后我们在代码中分别声明了两个隐式值，即 implicit val mInt=new MultiplicableInt 和 implicit val mStr=new MultiplicableString。当执行 multiply(5)时，编译器会在当前作用域内查找类型为 Multiplicable[T]的隐式值或隐式对象，这里显然为隐式值，隐式值 mInt 满足要求，因此 mInt 会被当作参数传递给 multiply 即相当于调用 multiply(5)(mInt)；执行 multiply("5")时也是类似的原理。

10.4.3 隐式参数使用常见问题

（1）当函数没有柯里化时，implicit 关键字会作用于函数参数列表中的所有参数。

```
//函数 product 中的 implicit 关键字会作用于 x:Double,y:Double，它们都是隐式参数
scala> def product(implicit x:Double,y:Double)=x*y
product: (implicit x: Double, implicit y: Double)Double

//定义一个隐式值
scala> implicit val d=2.55
d: Double = 2.55

//在未给定参数时，编译器会在当前作用域搜索相应的隐式值，在本例中为变量 d
//然后将该变量作为隐式函数参数，product 相当于 product(d,d)
scala> product
res1: Double = 6.5024999999999995
```

（2）隐式参数使用时要么全部指定，要么全不指定，不能只指定部分。

```
scala> def product(implicit x:Double,y:Double)=x*y
product: (implicit x: Double, implicit y: Double)Double

scala> implicit val d=3.0
d: Double = 3.0

//全部不指定
scala> product
res5: Double = 9.0

//不能只指定一个隐式参数
scala> product(3.0)
<console>:10: error: not enough arguments for method product: (implicit x: Double, implicit y: Double)Double.
 Unspecified value parameter y.
          product(3.0)

//全部指定
scala> product(3,3)
res3: Double = 9.0
```

（3）同类型的隐式值在当前作用域内只能出现一次。

```
scala> def product(implicit x:Double,y:Double)=x*y
product: (implicit x: Double, implicit y: Double)Double
```

```
scala> implicit val d=3.0
d: Double = 3.0

scala> product
res5: Double = 9.0

scala> implicit  val d1=4.0
d1: Double = 4.0

scala> product
<console>:13: error: ambiguous implicit values:
 both value d of type => Double
 and value d1 of type => Double
 match expected type Double
            product
            ^
```

（4）在定义隐式参数时，implicit 关键字只能出现在参数开头。

```
//隐式参数中的implicit关键字只能出现在参数开头：(implicit x:Double,y:Double)
scala> def product(implicit x:Double,y:Double)=x*y
product: (implicit x: Double, implicit y: Double)Double

//implicit关键字不在参数列表开头，则不合法：(x:Double,implicit y:Double)
scala> def product(x:Double,implicit y:Double)=x*y
<console>:1: error: identifier expected but 'implicit' found.
     def product(x:Double,implicit y:Double)=x*y
                          ^
```

（5）如果想达到函数中 def product(x:Double,implicit y:Double)只将参数 y 定义为隐式参数的目的，则需要对函数进行柯里化。

```
//函数柯里化后，只有函数参数 y:Double 为隐式参数
scala> def product(x:Double)(implicit y:Double)=x*y
product: (x: Double)(implicit y: Double)Double

scala> implicit val d=4.0
d: Double = 4.0

scala> product(3)
res2: Double = 12.0
```

（6）柯里化的函数 implicit 关键字只能作用于最后一个参数。

```
//implicit关键字作用于最后一个参数，合法
scala> def product(x:Double)(y:Double)(implicit z:Double)=x*y*z
product: (x: Double)(y: Double)(implicit z: Double)Double
```

```
//implicit 关键字作用于非最后一个参数，不合法
scala> def product(implicit x:Double)(y:Double)(z:Double)=x*y*z
<console>:1: error: '=' expected but '(' found.
      def product(implicit x:Double)(y:Double)(z:Double)=x*y*z

//implicit 关键字作用于非最后一个参数，不合法
scala> def product(x:Double)(implicit y:Double)(z:Double)=x*y*z
<console>:1: error: '=' expected but '(' found.
      def product(x:Double)(implicit y:Double)(z:Double)=x*y*z
```

（7）implicit 关键字在隐式参数中只能出现一次，对柯里化的函数也不例外。

```
//implicit 关键字只出现一次，合法
scala> def product(implicit x:Double,y:Double)=x*y

//implicit 关键字出现二次，不合法
scala> def product(implicit x:Double,implicit y:Double)
<console>:1: error: identifier expected but 'implicit' found.
      def product(implicit x:Double,implicit y:Double)

//对柯里化的函数也适用
scala> def product(implicit x:Double)(implicit y:Double)
<console>:1: error: '=' expected but '(' found.
      def product(implicit x:Double)(implicit y:Double)
```

（8）匿名函数不能使用隐式参数。

```
scala> val product=(x:Double,y:Double)=>x*y
product: (Double, Double) => Double = <function2>

//匿名函数(值函数)不能使用隐式参数
scala> val product=(implicit x:Double,y:Double)=>x*y
<console>:1: error: '=>' expected but ',' found.
      val product=(implicit x:Double,y:Double)=>x*y
```

（9）柯里化的函数如果有隐式参数，则不能使用其偏应用函数。

```
//柯里化函数 product
scala> def product(x:Double)(y:Double)=x*y
product: (x: Double)(y: Double)Double

//两个参数的偏应用函数
scala> val p1=product _
```

```
p1: Double => (Double => Double) = <function1>

scala> p1(3.0)(4.0)
res0: Double = 12.0

//一个参数的偏应用函数
scala> val p2=product(3.0) _
p2: Double => Double = <function1>

scala> p2(4.0)
res1: Double = 12.0

//将柯里化函数参数 y 声明为隐式参数
scala> def product(x:Double)(implicit y:Double)=x*y
product: (x: Double)(implicit y: Double)Double

//定义有隐式参数后，便不能使用其偏应用函数
scala> val p1=product _
<console>:8: error: could not find implicit value for parameter y: Double
       val p1=product _
```

10.5 隐式转换规则与问题

10.5.1 隐式转换的若干规则

（1）显式定义规则

在使用带有隐式参数的函数时，如果没有明确指定与参数类型相同的隐式值，编译器不会通过额外的隐式转换来满足该函数的要求，示例如下：

```
//带隐式参数的函数 product
scala> def product(implicit x:Double,y:Double)=x*y
product: (implicit x: Double, implicit y: Double)Double

//Int 类型的隐式变量 x
scala> implicit val x:Int=5
x: Int = 5

//直接调用函数 product，不指定参数时，编译器会在当前作用域内查找类型为 Double 的隐式值由于查找不到，所以会报错。
//虽然在当前作用域内存在 Int 类型的隐式变量 x，但它并不会通过隐式转换成 Double 类型来匹配
scala> product
<console>:10: error: could not find implicit value for parameter x: Double
```

```
                    product

//显式指定参数时，类型不匹配，则会自动进行隐式转换，将 Int 类型转换成 Double 类型来匹配函
数的参数类型
scala> product(5,5)
res1: Double = 25.0

//显式声明一个 Double 类型的隐式变量 x
scala> implicit val x:Double=5
x: Double = 5.0

//编译器会在当前作用域查找 Double 类型的隐式值，由于这里显式声明隐式变量 x，程序能够运行
scala> product
res2: Double = 25.0
```

在上述代码中，首先定义了一个带隐式参数的函数 def product(implicit x:Double,y:Double)=x*y，然后声明了一个类型为 Int 的隐式变量 x，不指定参数调用 product，编译器会在当前作用域内查找类型为 Double 的隐式值，由于不存在这样的隐式值，所以程序报错。这里重点要说明的是，直接显式地指定函数参数调用函数 product(5,5)能够得到正确结果，因为类型不匹配时会自动进行隐式转换，即将 Int 类型转换为 Double 类型，但不指定参数调用 product 函数时，隐式值 implicit val x:Int=5 并不会自动进行隐式转换，即将 Int 类型转换成 Double 类型来适配 product 函数的参数类型要求，必须通过 implicit val x:Double=5 这种显式定义的方式定义隐式值，这便是显式定义规则。

（2）作用域规则

不管是隐式值、隐式对象、隐式类或隐式转换函数，都必须在当前作用域内才能起作用，下面的代码给出了作用域规则的演示。

```
//将隐式转换函数放在对象 Utils 中
scala> object Utils{
  implicit def double2Int(x:Double)=x.toInt
}
//隐式转换函数不在当前作用域内，隐式转换失败
scala> val x:Int=2.55
<console>:10: error: type mismatch;
 found   : Double(2.55)
 required: Int
       val x:Int=2.55

//将隐式转换函数引入到当前作用域
scala> import Utils._
import Utils._
```

```
//隐式转换顺利执行
scala> val x:Int=2.55
x: Int = 2
```

代码中将 Double 类型到 Int 类型的隐式转换函数 double2Int 定义在单例对象 Uitl 中，执行 val x:Int=2.55 时会出错，这是因为隐式转换函数 double2Int 不在当前作用域内。通过 import Utils._ 将对象 Uitl 中的隐式转换函数 double2Int 引入到当前作用域，再执行 val x:Int=2.55 便能够使隐式转换顺利执行。对于隐式对象、隐式类、隐式值等也是一样的，只有在作用域范围内才能生效，这便是隐式转换的作用域规则。

（3）无歧义规则

在前面介绍隐式转换函数及隐式值等内容时，已经涉及到无歧义规则，所谓无歧义指的是不能存在多个隐式转换使代码合法，如代码中不应该存在两个隐式转换函数能够同时使某一类型转换成另外一种类型，也不应该存在类型相同的两个隐式值、主构造函数参数类型及成员方法等相同的两个隐式类。下面的代码给出的是隐式转换函数造成的歧义。

```
// double2Int, Double 类型到 Int 类型的隐式转换函数
scala> implicit def double2Int(x:Double)=x.toInt
double2Int: (x: Double)Int

// d2i, Double 类型到 Int 类型的隐式转换函数
scala> implicit def d2i(x:Double)=x.toInt
d2i: (x: Double)Int

//有歧义，因为存在两个 Double 类型到 Int 类型的隐式转换函数
scala> val x:Int=2.55
<console>:9: error: type mismatch;
 found   : Double(2.55)
 required: Int
Note that implicit conversions are not applicable because they are ambiguous:
 both method double2Int of type (x: Double)Int
 and method d2i of type (x: Double)Int
 are possible conversion functions from Double(2.55) to Int
       val x:Int=2.55
```

下面的代码演示的是隐式值造成的歧义性。

```
scala> def sum(implicit x:Int,y:Int)=x+y
sum: (implicit x: Int, implicit y: Int)Int

scala> implicit val x:Int=5
x: Int = 5
```

```
scala> implicit  val y:Int=6
y: Int = 6

scala> sum
<console>:11: error: ambiguous implicit values:
 both value x of type => Int
 and value y of type => Int
 match expected type Int
          sum
```

下面的代码演示的是隐式类造成的歧义：

```
scala> implicit class Dog(name:String){
  def bark=println(s"$name is barking")
}
defined class Dog

scala> implicit class D(name:String){
  def bark=println(s"$name is barking")
}
defined class D

scala> "Nacy".bark
<console>:12: error: type mismatch;
 found   : String("Nacy")
 required: ?{def bark: ?}
Note that implicit conversions are not applicable because they are ambiguous:
 both method Dog of type (name: String)Dog
 and method D of type (name: String)D
 are possible conversion functions from String("Nacy") to ?{def bark: ?}
          "Nacy".bark

<console>:12: error: value bark is not a member of String
          "Nacy".bark
```

（4）一次转换原则

隐式转换从源类型到目标类型只会经过一次转换，不会经过多次隐式转换达到。

```
//隐式类，参数为String类型
scala> implicit class Dog(val name:String){
  def getName=name
  def bark=println(s"$name is barking")
}
```

```
defined class Dog

//隐式类,参数为前面定义的 Dog 类
scala> implicit class SpecialDog(dog:Dog){
  def specialBark=println(dog.getName+" is barking Specially")
}
defined class SpecialDog

//String 类型没有 bark 方法,会尝试隐式转换成 Dog 类型,由于只通过一次隐式转换即可,合法
scala> "Nacy".bark
Nacy is barking

//下面的代码期望的是,String 类型隐式转换到 Dog 类型,然后 Dog 类型再隐式转换到 SpecialD
og 类型
//不合法,因为源类型到目标类型的隐式转换最多只允许一次
scala> "Nacy".specialBark
<console>:12: error: value specialBark is not a member of String
       "Nacy".specialBark
              ^
```

在上述代码中,分别定义了两个隐式类: implicit class Dog(val name:String 和 implicit class SpecialDog(dog:Dog),其中 Dog 类可以完成 String 类型到 Dog 类型的隐式转换,而 SpecialDog 类型可以完成 Dog 类型到 SpecialDog 类型的隐式转换。在执行代码"Nacy".bark 时,由于 String 类型不存在 bark 方法,此时会尝试进行隐式转换,隐式类 Dog 可以完成 String 类型到 Dog 类型的隐式转换,再调用 bark 方法,由于源类型(String 类型)到目标类型(Dog 类型)的转换只需要经过一次即可,因此它是合法的;在执行代码时,"Nacy".specialBark 之所以出错是因为不能期待编译器会先将 String 类型转换到 Dog 类型,再将 Dog 类型转换到 SpecialDog 类型,源类型(String 类型)到目标类型(SpecialDog 类型)的转换需要两步,不符合一次转换原则。

一次转换原则可以使源类型到目标类型的转换次数得到限制,使程序中的隐式转换能够在控制范围内,使代码的"迷幻"程度得到有效控制。

10.5.2 隐式转换需注意的问题

(1) 多次隐式转换问题

在前一小节介绍一次转换原则时,我们提到源类型到目标类型的隐式转换只会经过一次,但这并不意味着代码中不允许存在多次隐式转换,下面的代码给出了会发生多次隐式转换的例子。

```
/*
 * 多次隐式转换
 */
object Example10_4 extends  App{
```

```
class TestA{
  override def toString="this is TestA"
  def printA=println(this)
}

class TestB{
  override def toString="this is TestB"
  def printB(x:TestC)=println(x)
}

class TestC{
  override def toString="this is TestC"
  def printC=println(this)
}

//TestA 到 TestB 的隐式转换函数
implicit def A2B(x:TestA)={
  println("TestA is being converted to TestB")
  new TestB
}

//TestB 到 TestC 的隐式转换函数
implicit def B2C(x:TestB)={
  println("TestB is being converted to TestC")
  new TestC
}

val a=new TestA
//TestA 中不会存在 printB 方法,因此会尝试隐式转换,调用的是 implicit def A2B(x:TestA),这是第一次隐式转换
//在执行 printB 方法时,由于 def printB(x:TestC)接受的参数类型为 TestC,与实际不匹配,因此会尝试隐式转换
//调用的是 implicit def A2B(x:TestA),这是第二次隐式转换
a.printB(new TestB)
}
```

代码运行结果如下:

```
TestA is being converted to TestB
TestB is being converted to TestC
this is TestC
```

在代码中定义了 3 个类,它们分别是 TestA、TestB 及 TestC,需要注意的是 TestB 中的 def printB(x:TestC)成员方法参数为 TestC,打印输出 TestC 对象。然后,定义了两个隐式转换函数,分别是 implicit def A2B(x:TestA)和 implicit def B2C(x:TestB),隐式转换函数 A2B 完成

TestA 类型到 TestB 类型的隐式转换，而 B2C 完成 TestB 到 TestC 类型的隐式转换。最后，通过代码 val a=new TestA 创建相应的对象，代码最关键的地方是 a.printB(new TestB)，让我们对该方法的执行流程进行说明。

- 由于 TestA 中并未对 printB 方法下定义，因此编译器会尝试进行隐式转换。
- 调用 implicit def A2B(x:TestA)方法完成 TestA 类型到 TestB 类型的隐式转换（第一次隐式转换），打印输出 TestA is being converted to TestB。
- 调用 TestB 中定义的 def printB(x:TestC)方法，但 printB 只接受参数类型为 TestC 的对象，这里传入的参数与实际类型不匹配，因此编译器会尝试进行隐式转换。
- 调用 implicit def B2C(x:TestB)方法完成 TestB 到 TestC 类型的隐式转换（第二次隐式转换），打印输出 TestB is being converted to TestC。
- 最终执行 printB 方法，由于函数参数类型为隐式转换后的对象类型 TestC，因此打印输出 this is TestC。

可以看到代码 a.printB(new TestB)在执行时会发生多次隐式转换，但无论发生多少次隐式转换，一次转换原则仍然是不变的。

（2）是否要用隐式转换的问题

通过对前面隐式转换的介绍，不难看出隐式转换功能十分强大，但也增加了代码理解的复杂度，例如隐式类的使用使隐式转换隐藏得更深，同时多次隐式转换使代码执行逻辑变得非常复杂，这些都增加了代码的理解难度。那在实际开发时到底要不要用隐式转换呢？这是个很难回答的问题，部分隐式转换重度使用者认为：隐式转换虽然带来了代码的理解复杂度，但隐式转换可以快速地根据原有的类库扩展现有类的功能，只要很少的代码量便可以实现复杂的功能，代码量越少，开发人员出错的机率越低，相对于增加点代码的理解难度来说是值得的，理解难度只在于隐式转换的语法及作用规则；对于另外一部分 Scala 语言使用者来说，虽然 Scala 语言隐式转换无处不在，功能也很强大，但隐式转换为代码所带来的"迷幻"效果给程序的维护带来很大挑战，隐式转换不利于软件的持续升级和维护。

在实际开发过程中，是否使用隐式转换推荐遵循以下原则：

- 涉及多次隐式转换时，说服自己这样做的合理性，否则请对编写的代码进行重构。
- 在非必要的情况下尽量少用隐式转换，毕竟"迷幻"的代码理解起来比较困难。
- 如果是炫耀自己的 Scala 隐式转换技能，请大胆使用。

小 结

本章通过 Scala 默认提供的隐式转换函数说明隐式转换在 Scala 语言中是无处不在的，在此基础上引出隐式转换函数的定义与使用，介绍了隐式类与隐式对象的作用原理，详细介绍了隐式参数与隐式值的使用方法与原理，最后对隐式转换中的若干重要规则进行了介绍。在下一章中，我们将介绍 Scala 语言中的另外一项重要内容——类型参数。

第 11 章　类型参数

　　Scala 语言有着丰富的类型系统，本章将详细介绍 Scala 语言中的类型变量界定、视图界定、上下文界定、协变与逆变及高级类型等重要内容。

11.1 类与类型

在介绍 Scala 泛型之前，先对两个重要的概念进行介绍。Java 语言在引入泛型之前，对象的类（class）和类型（type）是一一对应的，如以下代码所示。

```java
package cn.scala.chapter11;
class Test{
}
public class JavaClassTypeDemo {
    public static void main(String[] args) {
        //获取String的类
        System.out.println("String.class="+String.class);
        //获取String对象的类型
        System.out.println("\"123\".getClass()="+"123".getClass());
        System.out.println(".............................................");
        //获取自定义Test的类
        System.out.println("Test.class="+Test.class);
        //获取自定义Test对象的类型
        System.out.println("new Test().getClass()="+new Test().getClass());
    }
}
```

上述代码运行结果如下：

```
String.class=class java.lang.String
"123".getClass()=class java.lang.String
.............................................
Test.class=class cn.scala.chapter11.Test
new Test().getClass()=class cn.scala.chapter11.Test
```

从运行结果可以看出，Java 中的类与类型是一一对应的。但随着泛型的引入，JVM 为了程序的兼容性采用了类型擦除，类与类型不再具有一致性，如以下代码所示。

```java
package cn.scala.chapter11;
import java.util.ArrayList;
import java.util.List;
class Test2<T>{
}
public class JavaClassTypeDemo2 {
    public static void main(String[] args) {
        //listStr对象的类型是List<String>
        List<String> listStr=new ArrayList<String>();
        //listInteger对象的类型是List<Integer>
        List<Integer> listInteger=new ArrayList<Integer>();
```

```java
        System.out.println(listStr.getClass());
        System.out.println(listInteger.getClass());

        //testStr 对象的类型是 Test2<String>
        Test2<String> testStr=new Test2<>();
        //testInteger 对象的类型是 Test2<Integer>
        Test2<Integer> testInteger=new Test2<>();

        System.out.println(testStr.getClass());
        System.out.println(testInteger.getClass());
    }
}
```

上述代码运行结果如下:

```
class java.util.ArrayList
class java.util.ArrayList
class cn.scala.chapter11.Test2
class cn.scala.chapter11.Test2
```

可以看到，虽然对象 listStr 与 listInteger 的类型、testStr 与 testInteger 的类型不一致，但它们对应的类却相同，也就是说类型比类更为具体，任何数据都是有类型的，例如 List<T>有 List<Integer>、List<String>等具体类型，但它们的类都为 List。Scala 中提供了更为丰富的类型系统，请看如下示例：

```scala
//引入下列包，便可以使用 typeOf 获取类型信息
scala> import scala.reflect.runtime.universe._
import scala.reflect.runtime.universe._

scala> class Test
defined class Test

scala> val a=new Test
a: Test = Test@15d227b

//获取类型信息
scala> typeOf[Test]
res1: reflect.runtime.universe.Type = Test

//获取类信息，相当于 Java 中的 Test.class
scala> classOf[Test]
res2: Class[Test] = class Test

//getClass 方法获取 Class[Test]的子类
scala> a.getClass
```

```
res3: Class[_ <: Test] = class Test

//listStr 的类型为 ArrayList[String]
scala> val listStr=new util.ArrayList[String]()
listStr: java.util.ArrayList[String] = []

//listInteger 的类型为 ArrayList[Integer]
scala> val listInteger=new util.ArrayList[Integer]()
listInteger: java.util.ArrayList[Integer] = []

//从下面两行代码结果可以看到，listInteger 的类与 listStr 的类都为 ArrayList，但其类型
不相同
scala> listInteger.getClass
res6: Class[_ <: java.util.ArrayList[Integer]] = class java.util.ArrayList

scala> listStr.getClass
res7: Class[_ <: java.util.ArrayList[String]] = class java.util.ArrayList

//下面的四行代码进一步说明了类与类型的不同
// List[Integer]与 List[String]类型不同
scala> typeOf[List[Integer]]
res4: reflect.runtime.universe.Type = java.util.List[java.lang.Integer]

scala> typeOf[List[String]]
res5: reflect.runtime.universe.Type = java.util.List[String]

// List[Integer]与 List[String]类都为 java.util.List
scala> classOf[List[Integer]]
res9: Class[java.util.List[Integer]] = interface java.util.List

scala> classOf[List[String]]
res10: Class[java.util.List[String]] = interface java.util.List
```

11.2 泛 型

Scala 泛型与 Java 泛型基本相同，只是 Java 泛型语法格式为<>，而 Scala 语法格式为[]，例如前一小节的 java.util.List，Java 泛型语法为 List<T>，在 Scala 中为 List[T]。

11.2.1 泛型类

Scala 泛型示例如下：

```scala
/**
  * Scala 泛型
  */
object ScalaExample11_1 extends  App{
  //Person[T]中的[T]为指定的泛型 T
  class Person[T](var name:T)
  class Student[T](name:T) extends Person(name)
  //在使用时将泛型参数具体化，这里为 String 类型
  println(new Student[String]("摇摆少年梦").name)

}
```

上面的例子是单个泛型参数的使用，泛型参数当然也可以有多个，例如：

```scala
/**
  * Scala 泛型：多个参数
  */
object ScalaExample11_2 extends  App{
  class Person[T](var name:T)
  //多个泛型参数
  class Student[T,S](name:T,var age:S) extends Person(name)

  //使用时指定 T 为 String 类型、S 为 Int 类型
  println(new Student[String,Int]("摇摆少年梦",18).name)

}
```

上述代码生成的字节码文件经过反编译后得到以下代码：

```
  E:\IntellijIDEAWorkspace\out\production\ScalaProject\cn\scala\chapter11>ja
vap -private ScalaExample11_2$Person.class
  Compiled from "chapter11.scala"
  public class cn.scala.chapter11.ScalaExample11_2$Person<T> {
    private T name;
    public T name();
    public void name_$eq(T);
    public cn.scala.chapter11.ScalaExample11_2$Person(T);
  }

  E:\IntellijIDEAWorkspace\out\production\ScalaProject\cn\scala\chapter11>ja
vap -private ScalaExample11_2$Student.class
```

```
Compiled from "chapter11.scala"
public class cn.scala.chapter11.ScalaExample11_2$Student<T, S> extends cn.s
cala.chapter11.ScalaExample11_2$Person<T> {
    private S age;
    public S age();
    public void age_$eq(S);
    public cn.scala.chapter11.ScalaExample11_2$Student(T, S);
}
```

通过观察反编译后的代码可以看出，Scala 泛型与 Java 泛型在使用时只是语法形式上有所不同。

11.2.2　泛型接口与泛型方法

除泛型类外，Scala 同 Java 一样，也可以有泛型接口和泛型方法，下列代码给出的是 java.util.Map 接口的定义。

```
// Java 泛型接口 Map<K,V>
public interface Map<K,V> {
    int size();
    boolean isEmpty();
    boolean containsKey(Object key);
    boolean containsValue(Object value);
    //Java 泛型方法
    V get(Object key);
    V put(K key, V value);
    V remove(Object key);
    void putAll(Map<? extends K, ? extends V> m);
    void clear();
    Set<K> keySet();
    Collection<V> values();
    Set<Map.Entry<K, V>> entrySet();
    interface Entry<K,V> {
        K getKey();
        V getValue();
        V setValue(V value);
        boolean equals(Object o);
        int hashCode();
    }
    boolean equals(Object o);
    int hashCode();
}
```

上面的代码给出的是 Java 泛型接口与泛型方法，Scala 中的泛型接口与泛型方法与 Java

语言中的泛型接口与泛型方法类似，下面给出的是 scala.collection.Map 类的代码：

```scala
//Scala 泛型接口，Scala 通过 trait 替代 Java interface 来实现泛型接口
trait Map[A, B]
  extends Iterable[(A, B)]
    with GenMap[A, B]
    with scala.collection.Map[A, B]
    with MapLike[A, B, Map[A, B]] {
  //泛型方法
  override def empty: Map[A, B] = Map.empty
  override def seq: Map[A, B] = this
  def withDefault(d: A => B): mutable.Map[A, B] = new Map.WithDefault[A, B](this, d)
  def withDefaultValue(d: B): mutable.Map[A, B] = new Map.WithDefault[A, B](this, x => d)
}
```

Scala 中通过 trait 来实现泛型接口，其泛型方法与 Java 语言也是类似的，只不过语法上略有差别。

11.2.3 类型通配符

前面我们提到，List<String>和 List<Integer>对象的类型是不一样的，但其类都为 List，在 Java 语言继承层次结构中，String 和 Integer 都是 Object 类的子类，但 List<String>和 List<Integer>与 List<Object>并不具备父子类的关系，下面的代码可以对此进行验证。

```
        List<String> listStr = new ArrayList<>();
        List<Integer> listInteger = new ArrayList<>();
        //下面的代码会报错
//Error:(13, 30) java: 不兼容的类型
//需要: java.util.List<java.lang.Object>
//找到:    java.util.List<java.lang.Integer>
        List<Object> listObj = listInteger;
```

由于这种关系导致在实际应用时会遇到不少问题，例如：

```java
import java.util.ArrayList;
import java.util.List;
public class JavaGenericDemo {
    public static void main(String[] args) {
        List<String> listStr = new ArrayList<>();
        List<Integer> listInteger = new ArrayList<>();
        //下面两行代码会出错，同样是因为 List<String>、List<Integer>与 List<Object>
//不是父子类关系
        printAll(listInteger);
        printAll(listStr);
```

```java
    }
    public static void  printAll(List<Object> listObj){
        //具体代码逻辑
    }
}
```

这一限制会给实际开发带来困难，在使用时不可能针对某个具体的类型来定义相应的 printAll 方法，因为这会造成方法的无限膨胀，这样的话显然背离了 Java 语言引入泛型的初衷，解决上述问题的方法是使用类型通配符。例如：

```java
import java.util.ArrayList;
import java.util.List;

public class JavaGenericDemo {
    public static void main(String[] args) {
        List<String> listStr = new ArrayList<>();
        List<Integer> listInteger = new ArrayList<>();
        printAll(listInteger);
        printAll(listStr);

    }
    //引入类型能配符？，List<?>可以视为 List<String>、List<Integer>所有类的父类
    public static void  printAll(List<?> list){
        //具体代码逻辑
    }
}
```

类型通配符还有两个特殊的用法，分别是类型通配符上限和类型通配符下限，类型通配符上限语法格式为 List<? extends Object>，其泛型参数可以接受 Object 子类（包括 Object 类本身），而类型通配符下限为 List<? super Integer>，其泛型参数可以接受所有 Integer 的父类（包括 Integer 类本身）。

在 Scala 语言当中，没有使用 Java 语言的这一语法，而是通过存在类型（Existential）来解决通配符问题，这么做的目的是解决 Java 通配符、原始类型擦除的兼容性问题，存在类型语法格式如下：

```scala
类[T] forSome {type T}
```

下面的例子说明了存在类型是如何使用的。

```scala
/**
 * Scala 存在类型
 */
object ScalaExample11_3 extends  App{
```

```
  val arrStr:Array[String]=Array("Hadoop","Hive","Spark")
  val arrInt:Array[Int]=Array(1,2,3)
  printAll(arrStr)
  printAll(arrInt)
  // 存在类型 Array[T] forSome {type T}
  def printAll(x: Array[T] forSome {type T})={
    for(i<- x){
      print(i+" ")
    }
    println()
  }
}
```

代码运行结果如下:

```
Hadoop Hive Spark
1 2 3
```

存在类型在使用时也可以被简化,方法 def printAll(x:Array[T] forSome {type T})可以简化为 def printAll(x:Array[_]),其中的"_"可以视为 Scala 风格的类型通配符,它是一种语法糖(Syntactic Sugar)。

```
/**
 * Scala 存在类型的简化使用
 */
object ScalaExample11_4 extends App{
  val arrStr:Array[String]=Array("Hadoop","Hive","Spark")
  val arrInt:Array[Int]=Array(1,2,3)
  printAll(arrStr)
  printAll(arrInt)
  //通过_简化设计,Array[_]与Array[T] forSome {type T}等价
  def printAll(x:Array[_])={
    for(i<- x){
      print(i+" ")
    }
    println()
  }
}
```

上面演示的是单个类型参数的存在类型,如果类型参数有多个,其语法格式如下:

```
类型[T,U,…] forSome {type T;type U;…}
```

下面是其使用示例:

```
//通配符表示
scala> def print(x:Map[_,_])=println(x)
```

```
print: (x: Map[_, _])Unit

//存在类型表示,与def print(x:Map[_,_])函数等价
scala> def print(x:Map[T,U] forSome {type T;type U})=println(x)
print: (x: Map[_, _])Unit
```

Scala 的类型系统比 Java 语言更复杂,同时也更灵活,Java 语言提供了类型通配符上限和类型通配符下限,Scala 语言也提供了类似的功能但还有其他更为复杂、灵活的用法。在本章后续几节中将会对 Scala 类型参数进行更为详细的介绍。

11.3 类型变量界定

在上一小节中提到 Java 语言中存在类型通配符上限这一特性,Scala 也提供了类似的功能,它便是类型变量界定(Type Variable Bound)。首先看下面的这段代码:

```
/**
 * 类型变量界定
 */
object ScalaExample11_5 extends App{
  //类 TypeVariableBound 编译通不过,因为泛型 T 在编译的时候不能确定其具体类型
  //即并不是所有的类中都存在 compareTo 方法
  class TypeVariableBound {
    def compare[T](first:T,second:T)={
      if (first.compareTo(second)>0)
        first
      else
        second
    }
  }

  val tvb=new TypeVariableBound
  println(tvb.compare("A", "B"))
}
```

上述代码中的类 TypeVariableBound 编译通不过,因为泛型 T 在编译的时候不能确定其具体类型,这是因为并不是所有的类都存在 compareTo 方法,但如果使用类型变量界定对泛型 T 的类型进行限定,也就是将 T 限定为 Comparable 的子类,此时 def compare[T](first:T,second:T) 方法中的参数 first 便存在 compareTo 方法,示例如下:

```
/**
 * 类型变量界定
 */
```

```
object ScalaExample11_5 extends App{
  class TypeVariableBound {
    //采用<:进行类型变量界定,该语法的意思是泛型T必须是实现了Comparable接口的类型
    def compare[T <: Comparable[T]](first:T,second:T)={
      if (first.compareTo(second)>0)
        first
      else
        second
    }
  }

  val tvb=new TypeVariableBound
  println(tvb.compare("A", "B"))
}
```

代码中 tvb.compare("A", "B")合法是因为 String 类型实现了 Comparable 接口。如果在程序中有开发人员自己定义的类,则需要手动实现 Comparable 接口,如以下代码所示:

```
/**
 * 类型变量界定
 */
object ScalaExample11_6 extends App{
  //声明Person类为case class且实现了Comparable接口
  case class Person(var name:String,var age:Int) extends Comparable[Person]{
    def compareTo(o:Person):Int={
      if (this.age>o.age) 1
      else if(this.age==o.age) 0
      else -1
    }
  }
  class TypeVariableBound {
    def compare[T <: Comparable[T]](first:T,second:T)={
      if (first.compareTo(second)>0)
        first
      else
        second
    }
  }

  val tvb=new TypeVariableBound
  println(tvb.compare("A", "B"))
  //此时下面这条语句是合法的,因为Person类实现了Comparable接口
  println(tvb.compare(Person("stephen",19), Person("john",20)))
```

}
```

通过代码可以看到，def compare[T <: Comparable[T]](first:T,second:T)方法由于参数使用了类型变量界定，只要 T 为 Comparable 的子类就是合法的。这里的类型变量界定作用在方法 compare 之上，除此之外类型变量界定还可作用于类中的泛型参数上，如以下代码：

```
/**
 * 作用于类型参数
 */
object ScalaExample11_7 extends  App{
  //定义Student类为case class，且泛型T的类型变量界定为AnyVal
  //在创建类时，所有处于AnyVal类继承层次结构的类都是合法的，如Int、Double等值类型
  case class Student[S,T <: AnyVal](var name:S,var hight:T)
  //下面这条语句是不合法的，因为String类型不属于AnyVal类层次结构
  // val S1=Student("john","170")
  //下面这两条语句都是合法的，因为Int,Long类型都是AnyVal
  val S2=Student("john",170.0)
  val S3=Student("john",170L)
}
```

case class Student[S,T <: AnyVal](var name:S,var hight:T)中的 T <: AnyVal 表示的是泛型 T 的最顶层类是 AnyVal，val S1=Student("john","170")，由于"170"是 String 类型，它不是 AnyVal 的子类，该语句不合法，因此像 T <: AnyVal 这种限制了 T 的最顶层类也被称为上界（Upper Bound）。与上界对应的是下界（Lower Bound），其语法格式如为[R >: T]，即泛型 R 必须是 T 的超类。

11.4　视图界定

上界和下界建立在 Scala 类继承层次结构基础之上，而本小节要讲的视图界定则可以跨越类继承层次结构，其实背后的实现原理是隐式转换。先看下面的代码：

```
/**
 * 不指定视图变量界定的情形
 */
object ScalaExample11_8 extends  App{
  //使用的是类型变量界定 S <: Comparable[S]
  case class Student[T,S <: Comparable[S]](var name:T,var height:S)

  //可以编译通过，因为String类型在Comparable继承层次体系
  val s= Student("john","170")
  //下面这条语句不合法，这是因为Int类型没有实现Comparable接口
  val s2= Student("john",170)
}
```
```

代码 val s= Student("john","170")因为 height 成员变量是 String 类型，而 String 类型实现了 Comparable 接口，因此该条语句可以编译通过；而 val s2= Student("john",170)语句不合法，是因为 Int 类型不属于 Comparable 类继承层次结果，但 RichInt 类型实现了 Comparable 接口，Int 类型可以经过隐式转换成 RichInt 类型，使用视图界定就可以使 val s2= Student("john",170)代码合法。示例如下：

```
/**
 * 指定为视图界定时
 */
object ScalaExample11_9 extends  App{
  //利用<%符号对泛型 S 进行限定，它的意思是 S 可以是 Comparable 类继承层次结构
  //中实现了 Comparable 接口的类也可以是能够经过隐式转换得到的类,该类实现了 Comparable 接口
  case class Student[T,S <% Comparable[S]](var name:T,var height:S)

  val s= Student("john","170")
  //下面这条语句在视图界定中是合法的因为 Int 类型此时会隐式转换为 RichInt 类，
  //而 RichInt 类实现了 Comparable 接口，属于 Comparable 继承层次结构
  val s2= Student("john",170)
}
```

代码中 S <% Comparable[S]使用了视图界定，它的意思是 S 可以是 Comparable 类继承层次结构中实现了 Comparable 接口的类，也可以是能够经过隐式转换后的类,该类实现了 Comparable 接口，代码 val s2= Student("john",170)合法是因为 Int 类型可以自动隐式转换为 RichInt 类，而 RichInt 类实现了 Comparable 接口。也就是说，类型变量界定要求类在类继承层次结构上，而视图界定不但可以在类继承层次结构上，还可以跨越类继承层次结构。

11.5　上下文界定

上下文界定（Context Bound）是 Scala 2.8 以后引入的一个新的语法特性，类型变量界定<:描述的是"is a"的关系，视图界定<%描述的是"can be seen as"的关系，而上下文界定描述的则是某种类型"has a"的关系。上下文界定的语言格式为 T：M，其中 M 是一个泛型，这种形式要求存在一个 M[T]类型的隐式值。下面的代码演示了上下文界定的使用方法：

```
/**
 * 上下文界定
 */
object ScalaExample11_10 extends  App{
  //PersonOrdering 混入了 Ordering 特质
  class PersonOrdering extends Ordering[Person]{
    override def compare(x:Person, y:Person):Int={
      if(x.name>y.name)
```

```
      1
    else
      -1
  }
}
case class Person(val name:String){
  println("正在构造对象:"+name)
}
//下面的代码定义了一个上下文界定
//它的意思是在对应作用域中,必须存在一个类型为Ordering[T]的隐式值,该隐式值可以作用于内部的方法
class Pair[T:Ordering](val first:T,val second:T){
  //smaller方法中有一个隐式参数,该隐式参数类型为Ordering[T]
  def smaller(implicit ord:Ordering[T])={
    if(ord.compare(first, second)>0)
      first
    else
      second
  }
}

//定义一个隐式值,它的类型为Ordering[Person]
implicit val p1=new PersonOrdering
val p=new Pair(Person("123"),Person("456"))
//不给函数指定参数,此时会查找一个隐式值,该隐式值类型为Ordering[Person]
//根据上下文界定的要求,p1正好满足要求
//因此它会作为smaller的隐式参数传入,从而调用ord.compare(first, second)方法进行比较
println(p.smaller)
}
```

代码 class Pair[T:Ordering](val first:T,val second:T)中的[T:Ordering]表示其泛型参数为一个上下文界定,在使用时要求代码中存在一个类型为Ordering[T]的隐式值,该隐式值在调用 def smaller(implicit ord:Ordering[T])方法时被传入,代码中的 implicit val p1=new PersonOrdering 便是我们定义的这个类型为Ordering[T]的隐式值。

有时候也希望 ord.compare(first, second)>0 的比较形式可以写为 first > second 这种直观的形式,此时可以省去 smaller 函数的隐式参数,并引入 Ordering 到 Ordered 的隐式转换,代码如下:

```
/**
 * 上下文界定
 */
object ScalaExample11_11 extends  App{
```

```scala
    //PersonOrdering混入了Ordering特质
    class PersonOrdering extends Ordering[Person]{
      override def compare(x:Person, y:Person):Int={
        if(x.name>y.name)
          1
        else
          -1
      }
    }
    case class Person(val name:String){
      println("正在构造对象:"+name)
    }
    //下面的代码定义了一个上下文界定
    //它的意思是在对应作用域中，必须存在一个类型为Ordering[T]的隐式值，该隐式值可以作用于内部的方法
    class Pair[T:Ordering](val first:T,val second:T){
      //smaller方法中有一个隐式参数，该隐式参数类型为Ordering[T]
      def smaller(implicit ord:Ordering[T])={
        if(ord.compare(first, second)>0)
          first
        else
          second
      }
    }

    //定义一个隐式值，它的类型为Ordering[Person]
    implicit val p1=new PersonOrdering
    val p=new Pair(Person("123"),Person("456"))
    //不给函数指定参数,此时会查找一个隐式值，该隐式值类型为Ordering[Person]
    //根据上下文界定的要求，p1正好满足要求
    //因此它会作为smaller的隐式参数传入，从而调用ord.compare(first, second)方法进行比较
    println(p.smaller)

}


/**
 * 上下文界定
 */
object ScalaExample11_10 extends  App{
  class PersonOrdering extends Ordering[Person]{
    override def compare(x:Person, y:Person):Int={
```

```
      if(x.name>y.name)
        1
      else
        -1
    }
  }
  case class Person(val name:String){
    println("正在构造对象:"+name)
  }

  class Pair[T:Ordering](val first:T,val second:T){
    //引入 odering 到 Ordered 的隐式转换
    //在查找作用域范围内的 Ordering[T]的隐式值
    //本例是 implicit val p1=new PersonOrdering
    //编译器看到比较方式是<的时候，会自动进行
    //隐式转换，转换成 Ordered，然后调用其中的<方法进行比较
    import Ordered.orderingToOrdered;
    def smaller={
      if(first<second)
        first
      else
        second
    }
  }

  implicit val p1=new PersonOrdering
  val p=new Pair(Person("123"),Person("456"))
  println(p.smaller)
}
```

11.6　多重界定

多重界定具有多种形式，例如：

- T:M:K //意味着在作用域中必须存在 M[T]、K[T]类型的隐式值。
- T<%M<%K //意味着在作用域中必须存在 T 到 M、T 到 K 的隐式转换。
- K>:T<:M //意味着 M 是 T 类型的超类，K 也是 T 类型的超类。

…

```
class A[T]
class B[T]
```

```
object MutilBound extends App{
 /**
  * 多重界定
  */
object ScalaExample11_12 extends  App{
  class A[T]
  class B[T]

  implicit val a=new A[String]
  implicit val b=new B[String]
  //多重上下文界定,必须存在两个隐式值,类型为A[T],B[T]类型
  //前面定义的两个隐式值a,b便是
  def test[T:A:B](x:T)=println(x)
  test("测试")

  implicit def t2A[T](x:T)=new A[T]
  implicit def t2B[T](x:T)=new B[T]
  //多重视图界定,必须存在T到A,T到B的隐式转换
  //前面我们定义的两个隐式转换函数就是
  def test2[T <% A[T] <% B[T]](x:T)=println(x)
  test2("测试 2")
}
```

代码 def test[T:A:B](x:T)=println(x)中使用了多重上下文界定,它的意思是在当前作用域中必须存在两个隐式值,类型分别为 A[T],B[T], 方法调用时如运行 test("测试")确定 T 为 String 类型,此时编译器会在当前作用域内查找 A[String]和 B[String]类型的隐式值,本示例中分别为 a 和 b。代码 def test2[T <% A[T] <% B[T]](x:T)=println(x)中使用了多重视图界定,它的意思是 T 必须为 A[T]的子类或存在 T 到 A[T]的隐式转换,T 必须为 B[T]的子类或存在 T 到 B[T]的隐式转换,在调用 test2("测试 2")时,确定 T 为 String 类型,由于代码中定义了两个隐式转换函数 t2A 及 t2B,可以满足条件,因此可以正常运行。

11.7 协变与逆变

在 Java 语言中,虽然 String 是 Object 的子类,但 List<String>并不是 List<Object>的子类。Java 语言不提供这种特性是有道理的,假设有 List<String> s1 和 List<Object> s2 两个变量,试想如果 List<String>是 List<Object>的子类这一条件成立,则变量 s1 可以赋值给变量 s2,那么再运行 s2.add(new Integer(123))也是合法的,但问题是 s1 引用的集合中的元素全部是 String 类型,这样显然会破坏类型安全。

Scala 比 Java 语言更为灵活,提供了非变(Nonvariance)、协变(covariance)与逆变

（Contravariance）三种特性。非变与 Java 语言提供的特性是一样的，如果 A 是 B 的子类，则 List[A]I 不是 List[B]的子类，示例代码如下：

```
/**
 * 非变
 */
object ScalaExample11_13 extends App{
  //定义自己的List类
  class List[T](val head: T, val tail: List[T])

  //编译报错 type mismatch; found :
  //cn.scala.chapter11.List[String] required:cn.scala.chapter11List[Any]
  //Note: String <: Any, but class List is invariant in type T.
  //You may wish to define T as +T instead. (SLS 4.5)
  val list:List[Any]= new List[String]("非变",null)
}
```

String 是 Any 的子类，代码 val list:List[Any]= new List[String]("非变",null)不合法的原因是类 class List[T](val head: T, val tail: List[T])是非变的，即 List[String]不是 List[Any]的子类。如果想要上述代码合法，则需要对 class List[T](val head: T, val tail: List[T])进行修改，将其定义为 class List[+T](val head: T, val tail: List[T])，具体代码如下：

```
/**
 * 协变
 */
object ScalaExample11_13 extends App{
  //用+标识泛型T，表示List类具有协变性
  class List[+T](val head: T, val tail: List[T])
  //List 泛型参数为协变之后，意味着List[String]也是List[Any]的子类
  val list:List[Any]= new List[String]("协变",null)
}
```

当 List 中的泛型参数指定为协变之后，在定义该类的方法时需要特别注意，见下列代码：

```
/**
 * 协变:逆变位置
 */
object ScalaExample11_13 extends App{
  class List[+T](val head: T, val tail: List[T]) {
    //下面的方法编译会出错
    //covariant type T occurs in contravariant position in type T of value newHead
    //编译器提示协变类型T出现在逆变的位置即泛型T定义为协变之后，泛型便不能直接应用于成员方法当中
    def prepend(newHead:T):List[T]=new List(newHead,this)
```

```
    }
    val list:List[Any]= new List[String]("摇摆少年梦",null)
}
```

上面的代码需要进行如下修改：

```
/**
  * 协变:逆变位置
  */
object ScalaExample11_13 extends  App{
  class List[+T](val head: T, val tail: List[T]) {
    //将函数也用泛型表示,因为是协变的,函数参数是个逆变点,参数类型为T的超类
    def prepend[U>:T](newHead:U):List[U]=new List(newHead,this)
    override def toString()=""+head
  }
  val list:List[Any]= new List[String]("协变",null)
}
```

为什么要这样做才合法呢？假设 def prepend(newHead:T):List[T]这么定义也是合法的，因为 Any 是 Strong 的父类，List[+T]的类型参数 T 又是协变的，因此 List[Any]是 List[String]的父类，定义两个变量分别是 List[Any] pAny=new List(123,null)和 List[String] pStr=new List("123",null)，便可以将 pStr 赋值给 pAny（里氏替换原则）：pAny=pStr，此时如果调用 pAny.prepend(246)是不合法的，因为子类 List[String]不接受非 String 类型的参数，即此时父类能做的，子类不能做，这显然违背了里氏替换原则。为满足里氏替换原则，必须将 def prepend(newHead:T):List[T]方法中的参数类型定义为 T 的超类，即定义为 def prepend[U>:T](newHead:U):List[U]才能满足。

除协变外，Scala 还提供逆变特性，它是指如果 A 是 B 的子类，则 List[B]是 List[A]的父类，因为加入类型参数后的关系与原来类 A、B 间的关系相反，从而被称之为逆变。逆变定义如下：

```
//定义 Person 的泛型参数为逆变-A
class Person[-A]{
  def test(x:A){}
}
```

具体示例如下：

```
/**
  * 逆变
  */
object ScalaExample11_13 extends App {

  class Person[-A] {
    def test(x: A): Unit = {
      println(x)
    }
  }
```

```
    val pAny = new Person[Any]
    //根据里氏替换原则,子类Person[Any]可以赋值给父类Person[String]
    val pStr: Person[String] = pAny
    pStr.test("Contravariance test----")
}
```

泛型参数为逆变的类中定义的方法也存在一个协变点,例如:

```
/**
 * 逆变:协变位置
 */
object ScalaExample11_13 extends App {

  class Person [-A]{
    def test(x:A): Unit ={
      println(x)
    }
    //下面这行代码会编译出错
    //contravariant type A occurs in covariant position in type   A of method test
    def test(x:A):A=null.asInstanceOf[A]
  }
}
```

def test(x:A): Unit 也可以用里氏替换原则来分析,同样声明两个变量:val pAnyRef=new Person[AnyRef]和 val pString=new Person[String],由于是逆变的,所以 Person[String]是 Person[AnyRef]的超类,pAnyRef 可以赋值给 pString,从而 pString 可以调用范围更广泛的函数参数(比如未赋值之前,pString.test("123")能为 String 类型,则 pAnyRef 赋值给 pString 之后,它可以调用 test(x:AnyRef)函数,使函数接受更广泛的参数类型。def test1:A 方法出错的原因也是因为子类函数处理范围更大,返回值类型也相应扩大,def test(x:A):A 方法返回值类型对应位置为协变点。

11.8 高级类型

Scala 语言除拥有基本的泛型、类型变量界定、视图界定、上下文界定等类型外,还拥有其他如单例类型、类型投影、类型别名、抽象类型、复合类型、函数类型、存在类型、中置类型及自身类型等高级类型,这些高级类型可以视为 Scala 语言的语法糖(Syntactic Sugar)[1],用于简化 Scala 程序的编写。自身类型、存在类型在前面章节中已有涉及,本节将介绍最常用

---

[1] https://www.zhihu.com/question/20651624

的高级类型如单例类型、类型投影、类型别名、抽象类型、复合类型及函数类型。

11.8.1 单例类型

单例类型是所有 Scala 对象包括单例对象都存在的一种类型，下面的代码给出的是单例对象的单例类型。

```
//定义一个单例对象
scala> object Dog
defined module Dog

//引入 typeOf 方法，该方法会返回获取到的具体类型
scala> import scala.reflect.runtime.universe.typeOf
import scala.reflect.runtime.universe.typeOf

//Dog.type 为 Dog 的单例类型
scala> typeOf[Dog.type ]
res0: reflect.runtime.universe.Type = Dog.type

//将 Dog 赋值给变量 x，变量 x 的类型为单例类型 Dog.type
scala> val x:Dog.type=Dog
x: Dog.type = Dog$@61366d
```

代码中的 object Dog 声明了一个单例对象，然后使用 import scala.reflect.runtime.universe.typeOf 将 typeOf 方法引入到当前作用域，typeOf 方法会返回具体的类型而不是泛型擦除后的类型。代码 typeOf[Dog.type]中的 Dog.type 为对象 Dog 的单例类型，typeOf 方法输出结果可以看到其类型为 Dog.type，代码 val x:Dog.type=Dog 将单例对象 Dog 赋值给变量 x，其类型为 Dog.type。

其他 Scala 对象也有其相应的单例类型，例如：

```
//定义普通类 Dog
scala> class Dog
defined class Dog

//创建一个普通的对象 dog1
scala> val dog1=new Dog
dog1: Dog = Dog@1e69724

//dog1 的单例类型为 dog1.type
scala> typeOf[dog1.type]
res3: reflect.runtime.universe.Type = dog1.type

//dog1.type 为 Dog 类的子类且它有唯一的对象实例 dog1
scala> val x1:dog1.type=dog1
```

```
x1: dog1.type = Dog@1e69724

//用于验证 dog1.type 为 Dog 类的子类这一事实
scala> typeOf[dog1.type] <:< typeOf[Dog]
res9: Boolean = true

//创建另外一个普通的对象 dog2
scala> val dog2=new Dog
dog2: Dog = Dog@141cc77

//其单例类型为 dog2.type
scala> typeOf[dog2.type]
res4: reflect.runtime.universe.Type = dog2.type

//dog2.type 为 Dog 类的子类且它仅有唯一的对象实例 dog2
scala> val x:dog2.type=dog2
x: dog2.type = Dog@141cc77

//dog1.type 只有唯一的对象实例 dog1,将 dog2 赋值给 dog1.type 类型的变量会产生类型不匹配
scala> val x:dog1.type=dog2
<console>:15: error: type mismatch;
 found   : dog2.type (with underlying type Dog)
 required: dog1.type
       val x:dog1.type=dog2
                       ^
```

代码 val dog1=new Dog 创建了一个普通的对象,该对象存在对应的单例类型 dog1.type,该类型是 Dog 类的一个子类并且它有且仅有一个对象实例,该对象就是 dog1,所以代码 val x1:dog1.type=dog1 将对象 dog1 赋值给类型为 dog1.type 的变量 x1 是合法的。对于变量 dog2 也是类似的, dog2 的单例类型是 dog2.type,它也有且仅有一个对象实例 dog2。代码 val x:dog1.type=dog2 出现类型不匹配的错误,是因为 dog1.type 只有一个对象实例 dog1, dog1.type 与 dog2.type 它们两者属于不同的类型,下面的代码给出了说明。

```
//dog1.type 与 dog2.type 属于不同的类型
scala> typeOf[dog1.type]==typeOf[dog2.type]
res5: Boolean = false
```

可以看到,typeOf[dog1.type]==typeOf[dog2.type]的返回值为 false,这证明了它们的类型是不同的。需要强调的是,对象的单例类型不同,但对象的类是相同的。

```
scala> dog1.getClass
res7: Class[_ <: Dog] = class Dog

//对象 dog1 与 dog2 对应的类相同
```

```
scala> dog1.getClass==dog2.getClass
res8: Boolean = true
```

这是显而易见的，因为 dog1 和 dog2 都是 Dog 类的实例。

那单例类型有什么作用呢？单例类型最常用的场景便是链式调用，先看一个不使用单例类型会出现什么情况的例子：

```
/**
 * 未使用单例类型时的链式调用
 */
object Example11_14 extends App{
  class Pet{
    private var name:String=null
    private var weight:Float=0.0f

    def setName(name:String)={
      this.name=name
      //返回调用对象，类型为 Pet
      this
    }
    def setWeight(weight:Float)={
      this.weight=weight
      //返回调用对象，类型为 Pet
      this
    }
    override  def toString=s"name=$name,age=$weight"
  }

  class Dog extends Pet{
    private var age:Int=0
    def setAge(age:Int)={
      this.age=age
      //返回调用对象，类型为 Dog
      this
    }
    override  def toString=super.toString+s",age=$age"
  }

  //代码能够顺利执行
  println(new Dog().setAge(2).setName("Nacy").setWeight(20.0f))

  //编译报错，Error:(33, 54) value setAge is not a member of cn.scala.chapter11.Example11_11.Pet
  //println(new Dog().setName("Nacy").setWeight(20.0f).setAge(2))
```

}

代码println(new Dog().setAge(2).setName("Nacy").setWeight(20.0f))能够编译通过并顺利执行，其执行步骤是：

（1）new Dog()创建了一个 Dog 对象，然后调用 Dog 类中的 setAge 成员方法，返回 Dog 对象。

（2）调用 setWeight 方法，因为 Dog 类继承了 Pet 中的 setWeight 方法，所以代码能够顺利执行，执行完成后返回的对象类型为 Pet。

（3）调用 setAge 方法，因为 Pet 中定义了 setAge 方法，所以代码能够顺利执行，执行完成后返回的对象类型为 Pet。代码 println(new Dog().setName("Nacy").setWeight(20.0f).setAge(2))编译通不过，是因为执行完 setName、setWeight 方法后其返回对象的类型为 Pet，当调用 setAge 方法时，由于 Pet 类中不存在该方法，从而报 value setAge is not a member of Pet 这一错误。

可以看到，方法的执行顺序对代码运行有很大的影响，然而在实际开发中不应该有这一限制，这时候便可以利用单例类型来解决这一问题，具体示例如下：

```scala
/**
 * 使用单例类型时的链式调用
 */
object Example11_15 extends App{

  class Pet{
    private var name:String=null
    private var weight:Float=0.0f
    //将函数返回值类型定义为单例类型
    def setName(name:String):this.type={
      this.name=name
      //返回调用对象，类型为 Pet
      this
    }

    //将函数返回值类型定义为单例类型
    def setWeight(weight:Float):this.type={
      this.weight=weight
      //返回调用对象，类型为 Pet
      this
    }
    override  def toString=s"name=$name,age=$weight"
  }

  class Dog extends Pet{
    private var age:Int=0
```

```
    //将函数返回值类型定义为单例类型
    def setAge(age:Int):this.type ={
      this.age=age
      //返回调用对象,类型为Dog
      this
    }
    override def toString=super.toString+s",age=$age"
}

//代码能够顺利执行
println(new Dog().setAge(2).setName("Nancy").setWeight(20.0f))

//同样能够顺利执行
println(new Dog().setName("Nancy").setWeight(20.0f).setAge(2))
}
```

Example11_12 中的代码与 Example11_11 中代码不同的地方在于 setAge、setName 及 setWeight 方法返回值类型全部定义为调用该方法的对象的单例类型:this.type,这样的话不同对象实例在执行对应方法时,返回的类型各不相同,都为调用该方法对象的单例类型。

11.8.2 类型投影

Scala 内部类同其成员变量一样,不同的对象实例对应的内部类是不同的,见下列代码。

```
scala> class Outter{
  val x:Int=0
  class Inner
}
defined class Outter

scala> val o1=new Outter
val o2=new Outter
o1: Outter = Outter@1ebe901
o2: Outter = Outter@7c1e7

scala> import scala.reflect.runtime.universe.typeOf
import scala.reflect.runtime.universe.typeOf

//o1.Inner 与 o2.Inner 为不同的类型
scala> typeOf[o1.Inner]==typeOf[o2.Inner]
res0: Boolean = false
```

代码中定义了一个类 class Outter,然后在该类中定义了一个内部类 class Inner,代码 val o1=new Outter 和 val o2=new Outter 定义了两个 Outter 对象,通过代码

typeOf[o1.Inner]==typeOf[o2.Inner]可以看到，o1.Inner 与 o2.Inner 类型不相同。类型不相同在使用内部类时会产生一些问题，例如：

```scala
/**
  * 类型投影
  */
object Example11_16 extends App{
  class Outter{
    val x:Int=0
    //定义 Outter 的成员方法 test，方法参数为内部类 Inner
    def test(i:Inner)=i
    class Inner
  }

  val o1=new Outter
  val o2=new Outter

  val inner2=new o2.Inner
  val inner1=new o1.Inner

  //编译通过
  o1.test(inner1)

  //编译出错
  //Error:(117, 12) type mismatch;
  //  found   : cn.scala.chapter11.Example11_13.o2.Inner
  //  required: cn.scala.chapter11.Example11_13.o1.Inner
  //  o1.test(inner2)
  //          ^
  o1.test(inner2)
}
```

在 Outter 类中定义了接受内部类作为成员方法 test 的参数，代码 o1.test(inner1)能够编译通过，但 o1.test(inner2)会编译报错，提示类型不匹配，其本质原因是 o2.Inner 与 o1.Inner 类型不同。在讲内部类的时候提到过，除了它是一个类之外，内部类同成员域和成员方法并没有区别。与成员域 x 一样，对象 o1.x 与对象 o2.x 本质上是不同的变量，它们在内存中的存储区域不同，而对 o1.Inner 与 o2.Inner 来说，内部类也是一样的，它们本质上是不同的类。

要使方法 def test(i:Inner)=i 可以同时接受 o1.Inner 与 o2.Inner 类型的对象，可以将方法定义为 def test(i:Outter#Inner)=I，函数参数中的 Outter#Inner 称为类型投影，其作用是将 o1.Inner 与 o2.Inner 两种不同的类型投影作为 Outter#Inner 的子类，从而使得 def test(i:Outter#Inner)方法可以接受 o1.Inner 与 o2.Inner 两种类型的对象，示例代码如下：

```
/**
```

* 类型投影
 */
```scala
object Example11_16 extends App{

  class Outter{
    val x:Int=0
    def test(i:Outter#Inner)=i
    class Inner
  }

  val o1=new Outter
  val o2=new Outter

  val inner2=new o2.Inner
  val inner1=new o1.Inner

  //编译通过
  o1.test(inner1)
  //编译通过
  o1.test(inner2)
}
```

11.8.3 类型别名

类型别名指的是给类取个另外的名字,在程序其他地方使用到这个类时便可以用这个别名来替代,示例如下：

```
//使用type关键字给类型取别名
scala> type JavaHashMap=java.util.HashMap[String,String]
defined type alias JavaHashMap

//使用别名创建HashMap对象
scala> val map=new JavaHashMap
map: java.util.HashMap[String,String] = {}

scala> map.put("Scala","2.10.4")
res1: String = null

scala> map
res2: java.util.HashMap[String,String] = {Scala=2.10.4}
```

代码 type JavaHashMap=java.util.HashMap[String,String] 使用 type 关键字给 java.util.HashMap[String,String] 取了个别名 JavaHashMap。当在程序中需要使用 java.util.HashMap[String,String] 时直接用 JavaHashMap 别名即可,如代码 val map=new

JavaHashMap 创建的便是 java.util.HashMap[String,String]类型的 HashMap。可以看到，类型别名可以简化程序设计，使代码更简洁。细心的读者会发现，类型别名与 import java.util.{HashMap=>JavaHashMap}重命名作用类似，虽然如此，但它们两者之前还是有区别的，例如：

```
//引入重命名不能带具体泛型参数类型
scala> import java.util.{HashMap[String,String]=>JavaHashMap}
<console>:1: error: '}' expected but '[' found.
       import java.util.{HashMap[String,String]=>JavaHashMap}
                                ^
scala> import java.util.{HashMap=>JavaHashMap}
import java.util.{HashMap=>JavaHashMap}

//类型别名必须指定具体泛型参数类型
scala> type JavaHashMap=java.util.HashMap
<console>:8: error: class HashMap takes type parameters
       type JavaHashMap=java.util.HashMap
                        ^
scala> type JavaHashMap=java.util.HashMap[String,String]
defined type alias JavaHashMap
```

通过上述代码可以看到，在使用类型别名时，如果类具有泛型参数则必须指定具体泛型类型，而在使用引入重命名时不能指定类具体的泛型类型。

11.8.4 抽象类型

抽象类型指的是在父类中类型不明确，而在子类中才能确定的类型。为说明抽象类型的概念及作用，先看下列示例：

```
/**
 * 抽象类型
 */
object Example11_17 extends App{

  class Food
  class Rice extends Food
  class Meat extends Food

  class Animal{
    def eat(f:Food)=f
  }
  class Human extends Animal{
    override def eat(f:Food)=f
  }
```

```
    class Tiger extends Animal{
      //方法参数只允许类型为Meat,但方法参数与父类Animal中的eat方法参数不同,不能override
      override  def eat(f:Meat)=f
    }
  }
```

代码中定义了食物类Food及其两个子类Rice和Meat,然后定义了Animal类及其两个子类Human和Tiger。在Animal类中定义了一个方法def eat(f:Food),其方法参数类型为Food,但在子类中需要对该方法进行重载,不过在Tiger这个类中,显然不能使用override def eat(f:Food)=f这种方式对父类的方法进行重构,因为如果这样的话,方法会接收类型为Rice的对象,显然老虎是不会吃米饭的。使用抽象类型可以解决这一问题,例如:

```
/**
 * 抽象类型
 */
object Example11_17 extends App{

  abstract class Food
  class Rice extends Food{
    override def toString="粮食"
  }
  class Meat extends Food{
    override def toString="肉"
  }

  class Animal{
    //定义了一抽象类型FoodType
    type FoodType
    //函数参数类型为FoodType
    def eat(f:FoodType)=f
  }
  class Human extends Animal{
    //子类中确定其具体类型为Food
    type FoodType=Food
    //函数参数类型为FoodType,因为已经具体化了,所以为Food
    override  def eat(f:FoodType)=f
  }
  class Tiger extends Animal{
    //子类中确定其具体类型为Meat
    type FoodType=Meat
    //函数参数类型为FoodType,因为已经具体化了,所以为Meat
    override  def eat(f:FoodType)=f
```

```
    }
    val human=new Human
    val tiger=new Tiger
    println("人可以吃："+human.eat(new Rice))
    println("人可以吃："+human.eat(new Meat))
    println("老虎只能吃："+tiger.eat(new Meat))
}
```

代码运行结果如下：

人可以吃：粮食
人可以吃：肉
老虎只能吃：肉

在 class Animal 中通过关键字 type 声明了一个抽象类型 FoodType：type FoodType，然后定义函数 def eat(f:FoodType)=f，函数参数用该抽象类型 FoodType 作为参数。在子类 class Human 中，将抽象类型具体化：type FoodType=Food，指定 FoodType 为 Food 类型，这样的话，方法 override def eat(f:FoodType) 相当于 override def eat(f: Food)。在子类 class Tiger 中，通过 type FoodType=Meat 将抽象类型具体化，指定其 FoodType 类型为 Meat 类型，方法 override def eat(f:FoodType) 相当于 override def eat(f: Meat)。这样，Human 的 eat 方法可以接受任意 Food 类及其子类对象，而 Tiger 的 eat 方法只能接受 Meat 类及子类对象。

11.8.5 复合类型

在定义类时，一个类常常会继承自一个类并混入若干 trait，例如：

```
final class RichInt(val self: Int) extends AnyVal with ScalaNumberProxy[Int] with RangedProxy[Int]
```

将 AnyVal with ScalaNumberProxy[Int] with RangedProxy[Int]看作一个整体，称这个整体为复合类型，下面的例子给出了复合类型的使用方式。

```
scala> class TestA
defined class TestA

scala> class TestB extends TestA with Cloneable
defined class TestB

//TestA with Cloneable 为复合类型，TestB 为该复合类型的子类
scala> def test(x:TestA with Cloneable)=println("ok")
test: (x: TestA with Cloneable)Unit

scala> test(new TestB)
ok
```

```
//TestC 同样为复合类型 TestA with Cloneable 的子类
scala> class TestC extends TestB
defined class TestC

scala> test(new TestC)
ok

//给复合类型取个别名
scala> type CompoundType=TestA with Cloneable
defined type alias CompoundType

scala> def test(x:CompoundType)=println("ok")
test: (x: CompoundType)Unit

scala> test(new TestB)
ok

scala> test(new TestC)
ok
```

代码中 class TestB extends TestA with Cloneable 定义了一个类 TestB，该类继承类 TestA 并混入 trait Cloneable，将 TestA with Cloneable 看成一个整体，这个整体便被称为复合类型。def test(x:TestA with Cloneable)=println("ok")定义了一个函数 test，该函数的参数类型为复合类型，它可以接受参数类型为所有继承该复合类型的类，代码中给出了传入对象为 TestB 类及其子类 TestC 的例子。type CompoundType=TestA with Cloneable 则给复合类型取了个别名，使用该别名可以简化程序设计。

11.8.6 函数类型

在 Scala 中函数也是有类型的，这是 Scala 语言提供的一个非常重要的语法，在高阶函数中常常会将函数对象作为一个函数的参数。例如：

```
//定义一个函数，该函数参数类型为函数类型
scala> def test(x:Double=>Int)=x(6.0)
test: (x: Double => Int)Int

scala> def double2int(x:Double)=x.toInt
double2int: (x: Double)Int

//函数作为参数进行传递
scala> test(double2int)
res5: Int = 6
```

代码 def test(x:Double=>Int)=x(6.0)定义了一个高阶函数 test，该函数的参数为函数类型

Double=>Int，表示传入的函数输入类型为 Double，返回值为 Int 类型，def double2int(x:Double)=x.toInt 定义了这样一个函数，然后作为参数传递给 test 函数 test(double2int)。事实上，Double=>Int 类型的函数有具体的类 Function1[Double,Int]与其对应，例如：

```
//通过类 Function1 创建函数对象
scala> val double2int=new Function1[Double,Int] {
  def apply(x:Double):Int=x.toInt
}
double2int: Double => Int = <function1>

scala> test(double2int)
res6: Int = 6
```

除此之外，Scala 中存在 FunctionN[T1,T2,…，TN+1]个函数类型，其中 N 的最大值为 22，之所以最大值为 22 是因为如果你设计的函数有超过 22 个参数类型，最大的问题可能在你的函数设计上，此时你需要对代码进行重构。下面再举个函数类型的例子，让大家对函数类型有个更深入的感受。

```
//创建函数类型对象 sum
scala> val sum=new Function2[Int,Int,Int] {
  def apply(x:Int,y:Int):Int=x+y
}
sum: (Int, Int) => Int = <function2>

scala> sum(4,6)
res7: Int = 10

//与创建函数类型对象 sum 的作用一样
scala> val sum=(x:Int,y:Int)=>x+y
sum: (Int, Int) => Int = <function2>

//与创建函数类型对象 sum 的作用一样
scala> def sum(x:Int,y:Int)=x+y
sum: (x: Int, y: Int)Int
```

代码

```
val sum=new Function2[Int,Int,Int] {
  def apply(x:Int,y:Int):Int=x+y
}
```

创建了一个函数类型对象 sum，该对象的类型为 Function2，它与函数字面量 val sum=(x:Int,y:Int)=>x+y 及函数 def sum(x:Int,y:Int)=x+y 是等价的，只不过后面两种函数定义方式更为简便和直观。

小 结

本章详细介绍了 Scala 语言中的类型系统，明确了 Scala 语言中类与类型的区别，给出了 Scala 中的泛型使用方法，着重介绍了 Scala 语言中的类型变量界定、视图界定、上下文界定、多重界定、协变、逆变及高级类型等。对视图界定、上下文界定等原理进行了深入分析，对 Scala 语言中的高级类型如单例类型、类型投影、类型别名、抽象类型、复合类型及函数类型的定义与使用进行了详细介绍。在下一章中，将详细介绍 Scala 语言中另外一项重要内容——并发编程。

第12章 Scala 并发编程基础

Scala 语言虽然支持 Java 语言中的多线程并发编程模式，但它同时又提供了 Actor 模型，使编写并发应用程序变得更简单的同时又避免了死锁等常见并发问题。在本章中将使用 Akka 框架进行并发应用程序编程。

12.1 Scala 并发编程简介

多核处理器的出现使并发编程（Concurrent Programming）成为开发人员必备的一项技能，许多现代编程语言都致力于解决并发编程的问题。并发编程虽然能够提高程序的性能，但传统并发编程的共享内存通信机制对开发人员的编程技能要求很高，需要开发人员通过自身的专业编程技能去避免死锁、互斥等待及竞争条件（Race Condition）等问题，熟悉 Java 语言并发编程的读者们对这些问题的理解会比较深刻，这些问题使得并发编程比顺序编程要困难得多。

Scala 语言并没有直接使用 Java 语言提供的并发编程库，而是通过 Actor 模型来解决 Java 并发编程中遇到的各种问题，为并发编程提供了更高级的抽象。

12.1.1 重要概念

（1）并发和并行

并发和并行从宏观来看，都是为进行多任务运行，但并发（Concurrency）和并行（Parallelism）两者之间是有区别的。并行是指两个或者两个以上的任务在同一时刻同时运行；而并发是指两个或两个以上的任务在同一时间段内运行，即一个时间段中有几个任务都处于已启动运行到运行完毕之间，这若干任务在同一 CPU 上运行但任一个时刻点上只有一个任务运行。图 12-1 给出了多核处理器下的现代操作系统进程和线程模型，图中进程 2 的线程 1 被调用到处理器的核 2 上运行、进程 3 的线程 1 被调度到处理器的核 3 上运行，进程 2 的线程 1 和进程 3 的线程 1 是并行的，它们可以同时运行，而进程 1 的线程 1 和线程 2 都调度到处理器的核 1 上运行，此外它们还共享进程 1 的内存空间，在运行时还面临着资源竞争包括 CPU、内存及其他如 IO 等，它们在同一时该只能运行一个，但在一段时间内都可以运行，因此进程 1 的线程 1 和线程 2 是并发执行的。

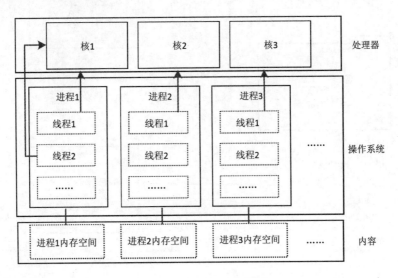

图 12-1 进程、线程模型

（2）横向扩展和纵向扩展

所谓纵向扩展（Scale Up）指的是增加程序的进程或线程数量，提高程序的并发性；而横向扩展（Scale Out）指的是程序可以扩展到其他机器上运行，即通过分布式系统来提高程序的并行度。传统的 Java 并发编程模型不容易进行纵向扩展，因此并发的线程数越多，程序行为便会变得很难理解和控制，当更多的线程加入到资源竞争中时，出现死锁等情况的概率会增加。横向扩展比纵向扩展困难更大，此时的程序变为分布式环境下的应用，情况更为复杂，对开发人员的要求更高。Scala 提供的 Actor 模型可以解决并发应用程序的横向扩展和纵向扩展问题，如图 12-2 和图 12-3 给出了基于 Actor 模型的横向扩展和纵向扩展示意图。

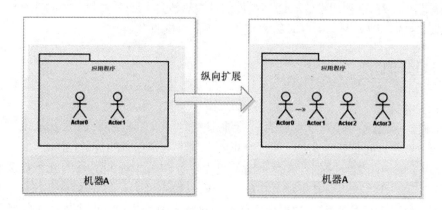

图 12-2　纵向扩展

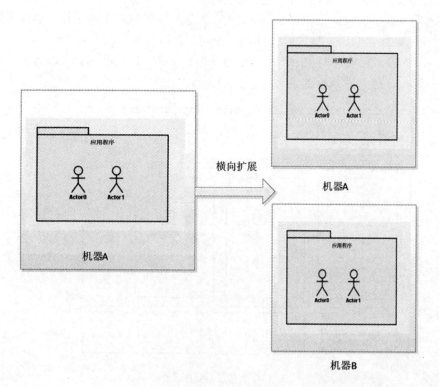

图 12-3　横向扩展

12.1.2 Actor 模型

在使用 Java 语言进行并发编程时，需要特别关注共享的数据结构及线程间的资源竞争容易导致死锁等问题，而 Actor 模型可以很大程度地解决这些问题。Actor 是一种基于事件（Event-Based）的轻量级线程，在使用 Actor 进行并发编程时只需要关注代码结构，而不需要过分关注数据结构，因此 Actor 最大限度地减少了数据的共享。Actor 由 3 个重要部分组成，它们是状态（State）、行为（Behavior）和邮箱（Mailbox），Actor 与 Actor 之间的交互通过消息发送来完成，Actor 模型如图 12-4 所示，状态指的是 Actor 对象的变量信息，它可以是 Actor 对象中的局部变量、占用的机器资源等，状态只会根据 Actor 接受的消息而改变，从而避免并发环境下的死锁等问题；行为指的是 Actor 的计算行为逻辑，它通过处理 Actor 接收的消息而改变 Actor 状态；邮箱建立起 Actor 间的连接，即 Actor 发送消息后，另外一个 Actor 将接收的消息放入到邮箱中等待后期处理，邮箱的内部实现是通过队列来实现的，队列可以是有界的（Bounded）也可以是无界的（Unbounded），有界队列实现的邮箱容量固定，无界队列实现的邮箱容量不受限制。

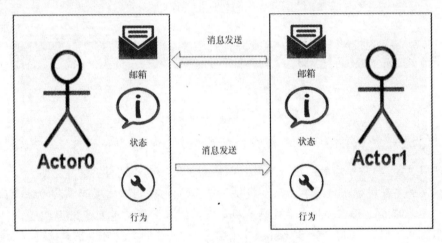

图 12-4　Actor 模型

不难看出，Actor 模型是对现实世界的高度抽象，它具有如下特点：

- Actor 之间使用消息传递机制进行通信，传递的消息使用的是不可变消息，Actor 之间并不共享数据结构，如果有数据共享则通过消息发送的方式进行。
- 各 Actor 都有对应的 mailbox，如果其他 Actor 向该 Actor 发送消息，消息将入队待后期处理。
- Actor 间的消息传递通过异步的方式进行，即消息的发送者发送完消息后不必等待回应便可以返回继续处理其他任务。

12.1.3 Akka 并发编程框架

Scala 语言中原生地支持 Actor 模型，只不过功能还不够强大，从 Scala 2.10 版本之后，Akka 框架成为 Scala 包的一部分，可以在程序中直接使用。Akka 是一个以 Actor 模型为基础构建的基于事件的并发编程框架，底层使用 Scala 语言实现，提供 Java 和 Scala 两种 API，它属于 LightBend 公司（原 Typesafe 公司）体系结构的一部分，如图 12-5 所示。

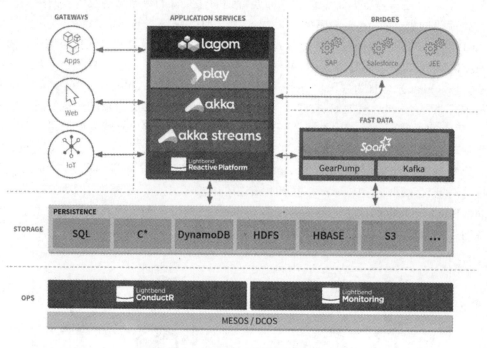

图 12-5　Lightbend 体系结构[1]

Akka 框架意在简化高并发、可扩展及分布式应用程序的设计，它具有如下优势：

- 使用 Akka 框架编写的应用程序既可以横向扩展，也可纵向扩展。
- 编写并发应用程序更简单，Akka 提供了更高的抽象，开发人员只需要专注于业务逻辑，而无需像 Java 语言那样需要处理底级语义如线程、锁及非阻塞 IO 等。
- 高容错。Akka 使用"let it crashes"机制，当 Actor 出错时可以快速恢复。
- 事件驱动的架构。Akka 中的 Actor 之间的通信采用异步消息发送，能够完美支持事件驱动。
- 位置透明。无论是 Actor 运行在本地机器还是远程机器上，对用户来说都是透明的，这极大地简化了多核处理器和分布式系统上的应用程序编程。
- 事务支持能力。支持软件事务内存（software transactional memory，STM），使 Actor 具有原子消息流的操作能力。

Akka 框架由下列十个组件构成：

（1）akka-actor　包括经典的 Actor、Typed Actors、IO Actor 等。

（2）akka-remote　远程 Actor。

（3）akka-testkit　测试 Actor 系统的工具箱。

---

[1] http://www.lightbend.com/products/lightbend-reactive-platform

（4）akka-kernel Akka 微内核，用于运行精简的微型应用程序服务器，无需运行于 Java 应用服务器上。

（5）akka-transactor Transactors 即支持事务的 actors，集成了 Scala STM。

（6）akka-agent 代理，同样集成了 Scala STM。

（7）akka-camel 集成 Apache Camel。

（8）akka-zeromq 集成 ZeroMQ 消息队列。

（9）akka-slf4j 支持 SLF4J 日志功能。

（10）akka-filebased-mailbox 支持基于文件的 mailbox。

本章后面的内容只涉及单台机器上的并发编程，主要为 akka-actor 中的类如 Actor、Untyped Actor 等，对于 akka-remote、akka-kernel 等其余组件的使用，可以参考 Akka 相关文献和官方文档。

12.2　Actor

12.2.1　定义 Actor

通过扩展 akka.actor.Actor 类并实现对应的 receive 方法进行 Actor 的定义，示例如下：

```
//定义自己的Actor,通过extends Actor并实现receive方法进行定义
 class StringActor extends Actor {
   val log = Logging(context.system, this)
   def receive = {
    case s:String   log.info("received message:"+s)
    case _          log.info("received unknown message")
   }
 }
```

上面的代码定义了一个名称为 StringActor 的 Actor，该 Actor 只处理 String 类型消息。StringActor 中实现了 receive 方法，在 akka.actor.Actor 类中，receive 方法具有如下定义：

```
type Receive = Actor.Receive
def receive: Receive
```

Actor.Receive 是一个偏函数：

```
type Receive = PartialFunction[Any, Unit]
```

该偏函数的输入类型可以是任意类型，返回值类型为 Unit。在实现的方法 receive 中通过标准的模式匹配进行输入消息的处理，例子中的 StringActor 中只处理 String 类型的输入。

12.2.2 创建 Actor

Actor 有 3 种创建方式：使用默认的构造函数创建 Actor；使用非默认的构造函数创建 Actor；通过隐式变量 context 创建 Actor。下面对这 3 种 Actor 创建方式分别进行介绍。

（1）使用默认的构造函数创建 Actor。

```scala
/**
 * 创建Actor：使用默认构造函数创建Actor
 */
object Example12_1 extends App{
  import akka.actor.Actor
  import akka.actor.Props
  import akka.event.Logging
  import akka.actor.ActorSystem

  //定义自己的Actor，通过extends Actor并实现receive方法进行定义
  class StringActor extends Actor {
    val log = Logging(context.system, this)
    def receive = {
      case s:String    log.info("received message:"+s)
      case _           log.info("received unknown message")
    }
  }

  //创建ActorSystem,ActorSystem为创建和查找Actor的入口
  //ActorSystem管理Actor共享配置信息如分发器(dispatchers)、部署（deployments）等
  val system = ActorSystem("StringSystem")

  //使用默认的构造函数创建Actor实例
  val stringActor = system.actorOf(Props[StringActor], name = "StringActor")

  //给stringActor发送字符串消息
  stringActor!"Creating Actors with default constructor"

  //关闭ActorSystem
  system.shutdown()
}
```

代码执行结果如下：

```
[INFO] [03/20/2016 19:40:17.529] [StringSystem-akka.actor.default-dispatcher-5] [akka://StringSystem/user/StringActor] received message:
Creating Actors with default constructor
```

代码 val system = ActorSystem("StringSystem")创建的是 ActorSystem 的实例，ActorSytem 是创建和查询 Actor 的入口，通过 ActorSystem 创建的对象为顶级 Actor，其 supervisor 为系统内部的守护 Actor。代码 val stringActor = system.actorOf(Props[StringActor], name = "StringActor")中的 actorOf 方法返回的是 ActorRef 的对象实例并赋值给变量 stringActor，通过 stringActor 便可以与创建的 Actor 进行交互，它是一种不可变的对象且与创建的 Actor 具有一一对应的关系，在实际使用时可以序列化并可以网络传输供远程调用，使用 actorOf 创建 Actor 时还可以指定创建的 Actor 名称，该名称用于标识创建的 Actor，不能与当前程序中的其他 Actor 有命名冲突。当不指定名称时 val stringActor = system.actorOf(Props[StringActor])，Akka 会自动帮我们命名：

```
[INFO] [03/20/2016 19:51:21.831] [StringSystem-akka.actor.default-dispatcher-2] [akka://StringSystem/user/$a] received message:
Creating Actors with default constructor
```

代码 stringActor!"Creating Actors with default constructor"为向 stringActor 发送字符串消息，该消息会被 receive 方法中的 case s:String ⇒ log.info("received message:"+s)语句处理。

（2）使用非默认的构造函数创建 Actor。

```
/**
 * 创建 Actor：使用非默认构造函数创建 Actor
 */
object Example12_2 extends App{
  import akka.actor.Actor
  import akka.actor.Props
  import akka.event.Logging
  import akka.actor.ActorSystem

  //定义自己的 Actor,通过 extends Actor 并实现 receive 方法进行定义
  class StringActor(var name:String) extends Actor {
    val log = Logging(context.system, this)
    def receive = {
      case s:String   log.info("received message:\n"+s)
      case _  ⇒ log.info("received unknown message")
    }
  }

  //创建 ActorSystem,ActorSystem 为创建和查找 Actor 的入口
  //ActorSystem 管理的 Actor 共享配置信息如分发器(dispatchers)、部署（deployments）等
  val system = ActorSystem("StringSystem")

  //使用非默认的构造函数创建 Actor 实例,注意这里是 Props(),而非 Props[]
  val stringActor = system.actorOf(Props(new StringActor("StringActor")),na
```

```
me="StringActor")

    //给 stringActor 发送字符串消息
    stringActor!"Creating Actors with non-default constructor"

    //关闭 ActorSystem
    system.shutdown()
}
```

代码运行结果如下:

```
[INFO] [03/20/2016 20:01:39.390] [StringSystem-akka.actor.default-dispatcher-5] [akka://StringSystem/user/StringActor] received message:
Creating Actors with non-default constructor
```

与 Example12_1 所不同的是,StringActor 定义了自己的主构造函数 class StringActor(var name:String),然后通过代码 val stringActor = system.actorOf(Props(new StringActor("StringActor")), name="StringActor")创建一个 Actor。其他代码逻辑与 Example12_1 类似,这里不再赘述。

(3)通过隐式变量 context 创建 Actor。

```
/**
 * 创建 Actor: 通过隐式变量 context 创建 Actor
 */
object Example12_3 extends App{
  import akka.actor.Actor
  import akka.actor.Props
  import akka.event.Logging
  import akka.actor.ActorSystem

  //定义自己的 Actor,通过 extends Actor 并实现 receive 方法进行定义
  class StringActor extends Actor {
    val log = Logging(context.system, this)

    def receive = {
      case s:String   log.info("received message:\n"+s)
      case _ ⇒ log.info("received unknown message")
    }
  }

  //再定义一个 Actor,在内部通过 context 创建 Actor
  class ContextActor extends Actor{
    val log = Logging(context.system, this)
    //通过 context 创建 StringActor
    var stringActor=context.actorOf(Props[StringActor],name="StringActor")
```

```
    def receive = {
      case s:String   log.info("received message:\n"+s);stringActor!s
      case _ ⇒ log.info("received unknown message")
    }
  }

  //创建ActorSystem,ActorSystem为创建和查找Actor的入口
  //ActorSystem管理的Actor共享配置信息如分发器(dispatchers)、部署(deployments)等
  val system = ActorSystem("StringSystem")

  //创建ContextActor
  val contextActor = system.actorOf(Props[ContextActor],name="ContextActor")

  //给contextActor发送字符串消息
  contextActor!"Creating Actors with implicit val context"

  //关闭ActorSystem
  system.shutdown()
}
```

代码运行结果如下:

```
[INFO] [03/20/2016 20:15:53.165] [StringSystem-akka.actor.default-dispatcher-4] [akka://StringSystem/user/ContextActor] received message:
Creating Actors with implicit val context
[INFO] [03/20/2016 20:15:53.169] [StringSystem-akka.actor.default-dispatcher-2] [akka://StringSystem/user/ContextActor/StringActor] received message:
Creating Actors with implicit val context
```

与 Example12_1 中的代码不同的是,在代码中新定义了一个 Actor 类 ContextActor,在 ContextActor 类中通过 var stringActor=context.actorOf(Props[StringActor],name="StringActor") 创建了 StringActor,ContextActor 类中的 receive 方法处理完发送过来的字符串消息后,又将消息发送给 StringActor 处理:case s:String ⇒ log.info("received message:\n"+s);stringActor!s。通过代码运行结果可以看到,与通过 ActorSystem 类中的 actorOf 所不同的是,使用 context 隐式变量创建的 Actor 会成为当前 Actor 的子 Actor,也就是 ContextActor 为 StringActor 的父 Actor(也称 Supervisor),日志输出[akka://StringSystem/user/ContextActor]、[akka://StringSystem/user/ContextActor/StringActor]证明了这一点。

在创建 Actor 特别是使用非默认构造函数创建 Actor 时,不能直接在代码中通过 new 关键字来创建 Actor 的实例,例如将 Example12_2 中的代码

```
val stringActor = system.actorOf(Props(new StringActor("StringActor")),name="StringActor")
```

改成如下所示：

```
val sa=new StringActor("StringActor")
val stringActor = system.actorOf(Props(sa),name="StringActor")
```

这样代码编译不会有问题，但实际执行时会报错，出错信息如下所示。

```
Exception in thread "main" 342f0891-4336-444f-916c-d06109688e54akka.actor.A
ctorInitializationException:
You cannot create an instance of [cn.scala.chapter12.Example12_2$StringActo
r] explicitly using the constructor (new).
You have to use one of the factory methods to create a new actor. Either use:
    'val actor = context.actorOf(Props[MyActor])'        (to create a supervi
sed child actor from within an actor), or
    'val actor = system.actorOf(Props(new MyActor(..)))' (to create a top le
vel actor from the ActorSystem)
    at akka.actor.ActorInitializationException$.apply(Actor.scala:171)
//省略其他无关错误信息
```

通过上述代码可以看到，Actor 的创建不能显式地使用 new 的方式创建，只能通过'val actor = context.actorOf(Props[MyActor])'或'val actor = system.actorOf(Props(new MyActor(..)))'这两种方式创建 Actor。

在 Example12_3 中可以看到[akka://StringSystem/user/ContextActor/StringActor]这样的输出信息，前面 Example12_1、Example12_2 代码也有类似的输出内容，它表示的是 Actor 的路径信息，对应 akka.actor. ActorPath 类，Actor 路径采用的是统一资源标识符（URI）的表示法，对应各字段含义如图 12-6 所示。

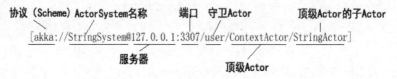

图 12-6　ActorPath 各字段含义

在例 Example12_3 中，代码 val system = ActorSystem("StringSystem")创建了一个 ActorSytem 并将其命名为 StringSystem，它对应的便是 ActorSytem 名称，由于 Example12_3 为本地机器上的 ActorSystem，所以服务器和端口默认没有显示，user 为每个 ActorSystem 都有的一个守卫 Actor（Guardian Actor），所有使用 system.actorOf 方法创建的 Actor 都是 user 的子 Actor，也称为应用程序的顶级 Actor，代码 system.actorOf(Props[ContextActor],name="ContextActor")创建的便是顶级 Actor ContextActor，而在 ContextActor 中通过 context.actorOf 方法创建的 Actor 便是该 Actor 的子 Actor，代码 var stringActor=context.actorOf(Props[StringActor], name="StringActor")创建的便是顶级 Actor ContextActor 的子 Actor StringActor。ActorSytem 正是通过这种方式构造了 Actor 层次结构模型，如图 12-7 所示，通过 ActorSystem 创建的 Actor 如 ContextActor，自动成为 user Actor 的子 Actor，而 user 则相应为 ContextActor 的 Supervisror

（监管者），在 ContextActor 中通过 context.actorOf 创建的 Actor 如 StringActor，将自动成为 ContextActor 的子 Actor，ContextActor 则自动成为 StringActor 的 Supervisor。Actor 这种层次模型对于 Akka 中的容错有着非常重要的影响，见本章 12.6 节的容错部分。

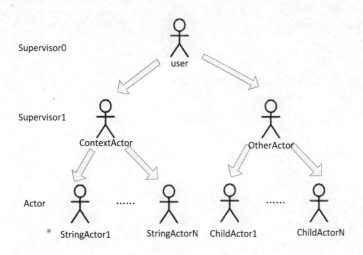

图 12-7　Actor 层次结构

12.2.3　消息处理

Akka 提供了两种消息模型：Fire-And-Forget 和 Send-And-Receive-Future。fire-and-forget 是一种单向消息发送模型，消息发送后可以立即返回，无需等待目标 Actor 返回结果，Akka 中使用!方法进行 fire-and-forget 消息发送，如 stringActor!"Creating Actors with implicit val context"，它的意思是当前发送方 Actor 向目标方 stringActor 发送字符串消息"Creating Actors with implicit val context"，发送完该消息后立即返回，而无需等待 stringActor 的返回，!还有个重载的方法 tell。Send-And-Receive-Future 则是一种双向消息发送模型，向目标 Actor 发送完消息后，返回一个 Future 作为后期可能的返回，当前发送方 Actor 将等待目标 Actor 的返回，Akka 中使用?方法进行 Send-And-Receive-Future 消息的发送，它也有一个重载的方法 ask。下面先给出 fire-and-forget 消息发送的代码演示：

```scala
/**
 * 消息处理：!(Fire-Forget)
 */
object Example12_4 extends App{
  import akka.actor.Actor
  import akka.actor.Props
  import akka.event.Logging
  import akka.actor.ActorSystem

  //定义几种不同的消息
  case class Start(var msg:String)
```

```scala
    case class Run(var msg:String)
    case class Stop(var msg:String)

    class ExampleActor extends Actor {
      val other = context.actorOf(Props[OtherActor], "OtherActor")
      val log = Logging(context.system, this)
      def receive={
        //使用 fire-and-forget 消息模型向 OtherActor 发送消息，隐式地传递 sender
        case Start(msg) => other ! msg
        //使用 fire-and-forget 消息模型向 OtherActor 发送消息，直接调用 tell 方法，显式指定 sender
        case Run(msg) => other.tell(msg, sender)
      }
    }

    class OtherActor extends Actor{
      val log = Logging(context.system, this)
      def receive ={
        case s:String=>log.info("received message:\n"+s)
        case _ => log.info("received unknown message")
      }
    }

    //创建 ActorSystem,ActorSystem 为创建和查找 Actor 的入口
    //ActorSystem 管理的 Actor 共享配置信息如分发器(dispatchers)、部署（deployments）等
    val system = ActorSystem("MessageProcessingSystem")

    //创建 ExampleActor
    val exampleActor = system.actorOf(Props[ExampleActor],name="ExampleActor")

    //使用 fire-and-forget 消息模型向 exampleActor 发送消息
    exampleActor!Run("Running")
    exampleActor!Start("Starting")

    //关闭 ActorSystem
    system.shutdown()
  }
```

代码运行结果如下：

[INFO] [03/20/2016 20:57:43.665] [MessageProcessingSystem-akka.actor.defaul

```
t-dispatcher-5] [akka://MessageProcessingSystem/user/ExampleActor/OtherActor] r
eceived message:
    Running
    [INFO] [03/20/2016 20:57:43.672] [MessageProcessingSystem-akka.actor.defaul
t-dispatcher-5] [akka://MessageProcessingSystem/user/ExampleActor/OtherActor] r
eceived message:
    Starting
```

在 ExampleActor 中，通过隐式变量 context 创建了 OtherActor 实例：val other = context.actorOf(Props[OtherActor], "OtherActor")，在 ExampleActor 的 receive 方法中，定义了处理两种不同类型消息的逻辑，具体代码为：

```
//使用 fire-and-forget 消息模型向 OtherActor 发送消息，隐式地传递 sender
case Start(msg) => other ! msg
//使用 fire-and-forget 消息模型向 OtherActor 发送消息，直接调用 tell 方法，显式指定 sender
case Run(msg) => other.tell(msg, sender)
```

处理 Start 类型的消息时，直接使用!进行消息发送，而处理 Run 类型的消息时，使用的是 tell 方法，可以看到使用 tell 方法需要显式地指定其 sender，而使用!进行消息发送则不需要，事实上!方法通过隐式值传入需要的 Sender，对比!与 tell 方法的定义便很容易理解：

```
//!方法的定义
def !(message: Any)(implicit sender: ActorRef = Actor.noSender): Unit
//tell 方法的定义
final def tell(msg: Any, sender: ActorRef): Unit = this.!(msg)(sender)
```

可以看到，tell 方法的实现依赖于!方法。如果在一个 Actor 当中使用!方法时，例如 ExampleActor 中使用的 other ! msg 向 OtherActor 发送消息，则 sender 隐式为 ExampleActor，如果不是在 Actor 中使用则默认为 Actor.noSender，即 sender 为 null。

理解了 Fire-And-Forget 消息模型后，接着对 Send-And-Receive-Future 消息模型进行介绍，下面的代码给出了其使用示例：

```
/**
 * 消息处理：?(Send-And-Receive-Future)
 */
object Example12_5 extends App{
    import akka.actor.Actor
    import akka.actor.Props
    import akka.event.Logging
    import akka.actor.ActorSystem
    import scala.concurrent.Future
    import akka.pattern.ask
    import akka.util.Timeout
    import scala.concurrent.duration._
```

```scala
import akka.pattern.pipe
import scala.concurrent.ExecutionContext.Implicits.global

//消息：个人基础信息
case class BasicInfo(id:Int,val name:String, age:Int)
//消息：个人兴趣信息
case class InterestInfo(id:Int,val interest:String)
//消息：完整个人信息
case class Person(basicInfo: BasicInfo,interestInfo: InterestInfo)


//基础信息对应Actor
class BasicInfoActor extends Actor{
  val log = Logging(context.system, this)
  def receive = {
    //处理发送而来的用户ID，然后将结果发送给sender（本例中对应CombineActor）
    case id:Int   log.info("id="+id);sender!new BasicInfo(id,"John",19)
    case _ => log.info("received unknown message")
  }
}


//兴趣爱好对应Actor
class InterestInfoActor extends Actor{
  val log = Logging(context.system, this)
  def receive = {
    //处理发送而来的用户ID，然后将结果发送给sender（本例中对应CombineActor）
    case id:Int  log.info("id="+id);sender!new InterestInfo(id,"足球")
    case _       log.info("received unknown message")
  }
}


//Person完整信息对应Actor
class PersonActor extends Actor{
  val log = Logging(context.system, this)
  def receive = {
    case person: Person =>log.info("Person="+person)
    case _        log.info("received unknown message")
  }
}



class CombineActor extends Actor{
  implicit val timeout = Timeout(5 seconds)
```

```
        val basicInfoActor = context.actorOf(Props[BasicInfoActor],name="BasicI
nfoActor")
        val interestInfoActor = context.actorOf(Props[InterestInfoActor],name="
InterestInfoActor")
        val personActor = context.actorOf(Props[PersonActor],name="PersonActor
")
        def receive = {
          case id: Int =>
            val combineResult: Future[Person] =
              for {
                //向 basicInfoActor 发送 Send-And-Receive-Future 消息，mapTo 方法将返回
结果映射为 BasicInfo 类型
                basicInfo <- ask(basicInfoActor, id).mapTo[BasicInfo]
                //向 interestInfoActor 发送 Send-And-Receive-Future 消息，mapTo 方法将
返回结果映射为 InterestInfo 类型
                interestInfo <- ask(interestInfoActor, id).mapTo[InterestInfo]
              } yield Person(basicInfo, interestInfo)

            //将 Future 结果发送给 PersonActor
          pipe(combineResult).to(personActor)
        }
      }

      val _system = ActorSystem("Send-And-Receive-Future")
      val combineActor = _system.actorOf(Props[CombineActor],name="CombineActor
")
      combineActor ! 12345
      Thread.sleep(5000)
      _system.shutdown

    }
```

代码运行结果如下：

```
[INFO] [03/20/2016 22:55:11.208] [Send-And-Receive-Future-akka.actor.defaul
t-dispatcher-3] [akka://Send-And-Receive-Future/user/CombineActor/BasicInfoAct
or] id=12345
    [INFO] [03/20/2016 22:55:11.220] [Send-And-Receive-Future-akka.actor.defaul
t-dispatcher-2] [akka://Send-And-Receive-Future/user/CombineActor/InterestInfo
Actor] id=12345
    [INFO] [03/20/2016 22:55:11.223] [Send-And-Receive-Future-akka.actor.defaul
t-dispatcher-4] [akka://Send-And-Receive-Future/user/CombineActor/PersonActor]
Person=Person(BasicInfo(12345,John,19),InterestInfo(12345,足球))
```

代码中定义了 3 种类型的消息，分别是个人基础信息 case class BasicInfo(id:Int,val

name:String, age:Int)、个人兴趣信息 case class InterestInfo(id:Int,val interest:String)以及完整个人信息 case class Person(basicInfo: BasicInfo,interestInfo: InterestInfo)，然后为这 3 种类型的消息定义了相应的 Actor 即 BasicInfoActor、InterestInfoActor 和 PersonActor。在 CombineActor 分别创建相应 Actor 的实例，receive 方法中使用 ask 向 BasicInfoActor、InterestInfoActor 发送 Send-And-Receive-Future 模型消息，BasicInfoActor、InterestInfoActor 中的 receive 方法接收到发送来的 Int 类型消息并分别使用!向 CombineActor 发送 BasicInfo、InterestInfo 消息，将结果保存在 Future[Person]中，然后通过代码 pipe(combineResult).to(personActor)将结果发送给 PersonActor。

12.2.4　Actor 的其他常用方法

Actor 中还有其他几个常用的方法，包括 preStart 方法、postStop 方法、unhandled 方法、preRestart 方法及 postRestart 方法，下面给出的是 preStart、postStop 及 unhandled 方法的使用示例。

```
/*
 * Actor 其他常用方法
 */
object Example12_6 extends App {
  class StringActor extends Actor {
    val log = Logging(context.system, this)

    //创建 Actor 时调用，在接受和处理消息前执行。主要用于 Actor 的初始化等工作
    override def preStart(): Unit = {
      log.info("preStart method in StringActor")
    }

    //Actor 停止时调用的方法
    override def postStop(): Unit = {
      log.info("postStop method in StringActor")
    }

    //有未能处理的消息时调用
    override def unhandled(message: Any): Unit = {
      log.info("unhandled method in StringActor")
      super.unhandled(message)
    }

    def receive = {
      case s:String    log.info("received message:\n"+s)
    }
```

```
    }

    val system = ActorSystem("StringSystem")

    //使用默认的构造函数创建 Actor 实例
    val stringActor = system.actorOf(Props[StringActor],name="StringActor")

    //给 stringActor 发送字符串消息
    stringActor!"Creating Actors with default constructor"

    //给 StringActor 发送整型数据,触发调用 unhandled 方法
    stringActor!123

    //关闭 ActorSystem
    system.shutdown()
  }
```

代码运行结果如下:

```
[INFO] [03/20/2016 23:21:19.603] [StringSystem-akka.actor.default-dispatcher-3] [akka://StringSystem/user/StringActor] preStart in StringActor
[INFO] [03/20/2016 23:21:19.608] [StringSystem-akka.actor.default-dispatcher-3] [akka://StringSystem/user/StringActor] received message: Creating Actors with default constructor
[INFO] [03/20/2016 23:21:19.608] [StringSystem-akka.actor.default-dispatcher-3] [akka://StringSystem/user/StringActor] unhandled in StringActor
[INFO] [03/20/2016 23:21:19.632] [StringSystem-akka.actor.default-dispatcher-5] [akka://StringSystem/user/StringActor] postStop in StringActor
```

通过代码输出可以看到,preStart 方法在创建 Actor 时会被调用,而 postStop 方法在 Actor 停止时调用,unhandled 方法会在 Actor 接收到未能处理的消息时触发执行。本例中的 StringActor 的 receive 方法只定义了 String 类型时的处理逻辑,因此当向 StringActor 给 StringActor 发送整型数据时,会触发执行 unhandled 方法。

12.2.5 停止 Actor

停止 Actor 的运行通常有 3 种方法:

- 通过 system.shutdown()停止所有 Actor 的运行。
- 通过发送 akka.actor.PoisonPill 消息给 Actor 停止 Actor,该消息像普通的消息一样加入到 mailbox 队列当中,Actor 处理该消息后便会停止运行。
- 通过调用 context.stop()方法停止 Actor 的运行,与调用 system.shutdown()方法会停止所有 Actor 运行所不同的是,context.stop()方法只能停止自身及其子 Actor 的运行。

下面的代码给出的是 context.stop()停止子 Actor 的运行示例。

```scala
/*
 * 停止Actor
 */
object Example12_7 extends App {
  import akka.actor.Actor
  import akka.actor.Props
  import akka.event.Logging
  import akka.actor.ActorSystem

  //定义自己的Actor，通过extends Actor并实现receive方法进行定义
  class StringActor extends Actor {
    val log = Logging(context.system, this)

    def receive = {
      case s:String   log.info("received message:\n"+s)
      case _          log.info("received unknown message")
    }

    override def postStop(): Unit = {
      log.info("postStop in StringActor")
    }
  }

  //再定义一个Actor，在内部通过context创建Actor
  class ContextActor extends Actor{
    val log = Logging(context.system, this)
    //通过context创建StringActor
    var stringActor=context.actorOf(Props[StringActor],name="StringActor")
    def receive = {
      case s:String   {
        log.info("received message:\n"+s)

        stringActor!s
        //停止StringActor
        context.stop(stringActor)
      }
      case _       log.info("received unknown message")
    }

    override def postStop(): Unit =  log.info("postStop in ContextActor")
  }
```

```
    //创建ActorSystem,ActorSystem为创建和查找Actor的入口
    //ActorSystem管理的Actor共享配置信息如分发器(dispatchers)、部署（deployments）
等
    val system = ActorSystem("StringSystem")


    //创建ContextActor
    val contextActor= system.actorOf(Props[ContextActor],name="ContextActor")

    //给contextActor发送字符串消息
    contextActor!"Creating Actors with implicit val context"

    //关闭ActorSystem
    //system.shutdown()
}
```

代码运行结果如下：

```
[INFO] [03/21/2016 07:15:52.524] [StringSystem-akka.actor.default-dispatcher-7] [akka://StringSystem/user/ContextActor] received message:
Creating Actors with implicit val context
[INFO] [03/21/2016 07:15:52.536] [StringSystem-akka.actor.default-dispatcher-2] [akka://StringSystem/user/ContextActor/StringActor] received message:
Creating Actors with implicit val context
[INFO] [03/21/2016 07:15:52.540] [StringSystem-akka.actor.default-dispatcher-7] [akka://StringSystem/user/ContextActor/StringActor] postStop in StringActor
```

在ContextActor中的receive方法中调用context.stop(stringActor)停止StringActor的运行，在处理完消息后，便停止运行，然后调用postStop()方法。

下面的代码给出的是发送akka.actor.PoisonPill消息停止Actor的示例。

```
/*
 * 停止Actor：使用PosionPill
 */
object Example12_8 extends App {

  import akka.actor.Actor
  import akka.actor.Props
  import akka.event.Logging
  import akka.actor.ActorSystem
  import scala.concurrent.Future
  import scala.concurrent.duration._
```

```scala
//定义自己的Actor，通过extends Actor并实现receive方法进行定义
class StringActor extends Actor {
  val log = Logging(context.system, this)

  def receive = {
    case s:String   log.info("received message:\n"+s)
    case _          log.info("received unknown message")
  }

  override def postStop(): Unit = {
    log.info("postStop in StringActor")
  }
}

//再定义一个Actor，在内部通过context创建Actor
class ContextActor extends Actor{
  val log = Logging(context.system, this)
  //通过context创建StringActor
  var stringActor=context.actorOf(Props[StringActor],name="StringActor")
  def receive = {
    case s:String   {
      log.info("received message:\n"+s)

      stringActor!s
      //停止StringActor
      context.stop(stringActor)
    }
    case _    log.info("received unknown message")
  }

  override def postStop(): Unit =  log.info("postStop in ContextActor")
}


//创建ActorSystem,ActorSystem为创建和查找Actor的入口
//ActorSystem管理的Actor共享配置信息如分发器(dispatchers)、部署（deployments）等
val system = ActorSystem("StringSystem")


//创建ContextActor
val contextActor = system.actorOf(Props[ContextActor],name="ContextActor")
```

```
    contextActor!"Creating Actors with implicit val context"

    //发送 PoisonPill 消息,停止 Actor
    contextActor!PoisonPill

    //关闭 ActorSystem
    //system.shutdown()
}
```

代码运行结果如下:

```
[INFO] [03/21/2016 09:45:11.847] [StringSystem-akka.actor.default-dispatcher-5] [akka://StringSystem/user/ContextActor] received message:
Creating Actors with implicit val context
[INFO] [03/21/2016 09:45:11.861] [StringSystem-akka.actor.default-dispatcher-2] [akka://StringSystem/user/ContextActor/StringActor] received message:
Creating Actors with implicit val context
[INFO] [03/21/2016 09:45:11.861] [StringSystem-akka.actor.default-dispatcher-4] [akka://StringSystem/user/ContextActor/StringActor] postStop in StringActor
[INFO] [03/21/2016 09:45:11.891] [StringSystem-akka.actor.default-dispatcher-5] [akka://StringSystem/user/ContextActor] postStop in ContextActor
```

上述代码中的 contextActor!PoisonPill 向 contextActor 发送 PoisonPill 消息达到停止 contextActor 运行的目的。

12.3 Typed Actor

12.3.1 Typed Actor 定义

Akka 中的 Typed Actor 是 Active Objects 设计模式的实现,Active Objects 模式将方法的执行和方法的调用进行解耦合,从而为程序引入并发性。Typed Actor 由公用的接口和对应实现两部分构成,其背后深层次的实现原理便是经典的代理模式,即通过使用 JDK 中的动态代理来实现,在调用接口的方法时自动分发到实现接口的对象上。Typed Actor 的定义[1]如下所示:

```
trait Squarer {
    //fire-and-forget 消息
    def squareDontCare(i: Int): Unit
    //非阻塞 send-request-reply 消息
    def square(i: Int): Future[Int]
    //阻塞式的 send-request-reply 消息
```

---

1 http://doc.akka.io/docs/akka/snapshot/scala/typed-actors.html

```
  def squareNowPlease(i: Int): Option[Int]
  //阻塞式的send-request-reply消息
  def squareNow(i: Int): Int
}

class SquarerImpl(val name: String) extends Squarer {
  def this() = this("SquarerImpl")

  def squareDontCare(i: Int): Unit = i * i
  def square(i: Int): Future[Int] = Promise.successful(i * i).future
  def squareNowPlease(i: Int): Option[Int] = Some(i * i)
  def squareNow(i: Int): Int = i * i
}
```

trait Squarer 中定义了 4 个方法：

（1）def squareDontCare(i: Int): Unit 方法。返回值类型为 Unit，它类似于 Untyped Actor 中的 fire-and-forget 消息发送模型，即！和 tell 方法调用。

（2）def square(i: Int): Future[Int]。返回值类型为 Future[Int]，它类似于 Untyped Actor 中的 send-request-reply 消息发送模型，即?和 ask 方法调用，此种调用是非阻塞的。

（3）def squareNowPlease(i: Int): Option[Int]。返回值类型为 Option[Int]（Option 类可以是 scala.Option[_]也可以是 akka.japi.Option<?>），它也类似于 Untyped Actor 中的 send-request-reply 消息发送模型，只不过它是阻塞式的调用。当在限定的时间内无返回值时，其返回结果为 None，否则为 Option[Int]。

（4）def squareNow(i: Int): Int。返回值类型为 Int，其作用与 def squareNowPlease(i: Int): Option[Int] 一样，只不过在超时情况下无返回值或出现其他异常时会抛出 java.util.concurrent.TimeoutException 异常。

在类 SquarerImpl 中实现了 trait Squarer 中定义的 4 个方法，对于 def square(i: Int): Future[Int]方法，使用 Promise.successful(i * i)传入一个表达式并阻塞执行直至运行完成，只不过方法的调用与执行时已经被解耦为异步执行。

12.3.2 创建 Typed Actor

通过下列代码创建 Typed Actor：

```
//直接通过默认的构造函数创建 Typed Actor
val mySquarer: Squarer =TypedActor(system).typedActorOf(TypedProps[SquarerImpl]())
//直接通过默认的构造函数创建 Typed Actor 并指定 Typed Actor 名称
val mySquarer: Squarer =TypedActor(system).typedActorOf(TypedProps[SquarerImpl](),"mySquarer")
//通过非默认的构造函数创建 Typed Actor 并指定 Typed Actor 名称
```

```
val otherSquarer: Squarer = TypedActor(system).typedActorOf(TypedProps(clas
sOf[Squarer],new SquarerImpl("SquarerImpl")), "otherSquarer")
```

上面代码演示的是使用构造函数和非默认构造函数创建 Typed Actor，其中 Squarer 为代理的类型，SquarerImpl 为具体实现的类型。

12.3.3 消息发送

示例代码如下：

```
//fire-forget 消息发送
 mySquarer.squareDontCare(10)

 //send-request-reply 消息发送
 val oSquare = mySquarer.squareNowPlease(10)

 val iSquare = mySquarer.squareNow(10)

 //Request-reply-with-future 消息发送
 val fSquare = mySquarer.square(10)
 val result = Await.result(fSquare, 5 second)
```

代码 mySquarer.squareDontCare(10)是单向消息发送，方法将在另外一个线程上异步地执行；val oSquare = mySquarer.squareNowPlease(10)和 val iSquare = mySquarer.squareNow(10)为 Request-reply 消息发送，在特定时间内以阻塞的方式执行，对于.squareNowPlease(10)方法如果在对应时间内没有返回结果则返回值为 None，否则返回值为 Option[Int]类型，对于 squareNow(10) 方法如果在对应时间内无返回值则会抛出异常 java.util.concurrent.TimeoutException，否则返回 Int 类型值；val fSquare = mySquarer.square(10) 为 Request-reply-with-future 式的消息发送，以非阻塞的方式执行，可以通过 val result = Await.result(fSquare, 5 second)获取执行结果。完整代码如下：

```
/*
 * Typed Actor
 */
object Example12_9 extends App {

  import akka.event.Logging
  import scala.concurrent.{ Promise, Future }
  import akka.actor.{ TypedActor, TypedProps }
  import scala.concurrent.duration._

  trait Squarer {
    //fire-and-forget 消息
    def squareDontCare(i: Int): Unit
```

```scala
  //非阻塞send-request-reply消息
  def square(i: Int): Future[Int]
  //阻塞式的send-request-reply消息
  def squareNowPlease(i: Int): Option[Int]
  //阻塞式的send-request-reply消息
  def squareNow(i: Int): Int
}

class SquarerImpl(val name: String) extends Squarer {
  def this() = this("SquarerImpl")

  def squareDontCare(i: Int): Unit = i * i
  def square(i: Int): Future[Int] = Promise.successful(i * i).future
  def squareNowPlease(i: Int): Option[Int] = Some(i * i)
  def squareNow(i: Int): Int = i * i
}

val system = ActorSystem("TypedActorSystem")
val log = Logging(system, this.getClass)

//使用默认构造函数创建Typed Actor
val mySquarer: Squarer =
  TypedActor(system).typedActorOf(TypedProps[SquarerImpl](),"mySquarer")

//使用非默认构造函数创建Typed Actor
  val otherSquarer: Squarer =
    TypedActor(system).typedActorOf(TypedProps(classOf[Squarer],
      new SquarerImpl("SquarerImpl")), "otherSquarer")


//fire-forget 消息发送
mySquarer.squareDontCare(10)

//send-request-reply 消息发送
val oSquare = mySquarer.squareNowPlease(10)

log.info("oSquare="+oSquare)

val iSquare = mySquarer.squareNow(10)
log.info("iSquare="+iSquare)

//Request-reply-with-future 消息发送
val fSquare = mySquarer.square(10)
```

```
    val result = Await.result(fSquare, 5 second)

    log.info("fSquare="+result)

    system.shutdown()
}
```

代码运行结果如下:

```
[INFO] [03/21/2016 21:15:50.592] [main] [Example12_9$(akka://TypedActorSystem)] oSquare=Some(100)
[INFO] [03/21/2016 21:15:50.649] [main] [Example12_9$(akka://TypedActorSystem)] iSquare=100
[INFO] [03/21/2016 21:15:50.649] [main] [Example12_9$(akka://TypedActorSystem)] fSquare=100
```

12.3.4 停止运行 Typed Actor

当 Typed Actor 不再需要时要将其停止,有 3 种方法停止 Typed Actor 的运行:

- 通过 system.shutdown()停止 ActorSystem 中所有的 Typed Actor。
- 调用 TypedActor(system).stop(mySquarer)停止指定的 Typed Actor。
- 调用 TypedActor(system).poisonPill(otherSquarer)停止指定的 Typed Actor。

具体示例代码如下:

```
/*
 * 停止 Typed Actor
 */
object Example12_10 extends App {

  import akka.event.Logging
  import scala.concurrent.{ Promise, Future }
  import akka.actor.{ TypedActor, TypedProps }
  import scala.concurrent.duration._

  trait Squarer {
    //fire-and-forget 消息
    def squareDontCare(i: Int): Unit
    //非阻塞 send-request-reply 消息
    def square(i: Int): Future[Int]
    //阻塞式的 send-request-reply 消息
    def squareNowPlease(i: Int): Option[Int]
    //阻塞式的 send-request-reply 消息
    def squareNow(i: Int): Int
```

```scala
    }

    //混入 PostStop 和 PreStart
    class SquarerImpl(val name: String) extends  Squarer with PostStop with PreStart {
      import TypedActor.context
      val log = Logging(context.system,TypedActor.self.getClass())
      def this() = this("SquarerImpl")

      def squareDontCare(i: Int): Unit = i * i
      def square(i: Int): Future[Int] = Promise.successful(i * i).future
      def squareNowPlease(i: Int): Option[Int] = Some(i * i)
      def squareNow(i: Int): Int = i * i

      def postStop(): Unit={
        log.info ("TypedActor Stopped")
      }
      def preStart(): Unit={
        log.info ("TypedActor Started")
      }
    }


    val system = ActorSystem("TypedActorSystem")
    val log = Logging(system, this.getClass)

    //使用默认构造函数创建 Typed Actor
    val mySquarer: Squarer =
      TypedActor(system).typedActorOf(TypedProps[SquarerImpl](),"mySquarer")



    //使用非默认构造函数创建 Typed Actor
    val otherSquarer: Squarer =
      TypedActor(system).typedActorOf(TypedProps(classOf[Squarer],
        new SquarerImpl("SquarerImpl")), "otherSquarer")

    //Request-reply-with-future 消息发送
    val fSquare = mySquarer.square(10)
    val result = Await.result(fSquare, 5 second)
    log.info("fSquare="+result)

    //调用 poisonPill 方法停止 Actor 运行
```

```
    TypedActor(system).poisonPill(otherSquarer)

    //调用 stop 方法停止 Actor 运行
    TypedActor(system).stop(mySquarer)

    //system.shutdown()
}
```

代码运行结果如下:

```
[INFO] [03/21/2016 22:41:51.119] [TypedActorSystem-akka.actor.default-dispa
tcher-2] [$Proxy0(akka://TypedActorSystem)] TypedActor Started
[INFO] [03/21/2016 22:41:51.123] [TypedActorSystem-akka.actor.default-dispa
tcher-2] [$Proxy1(akka://TypedActorSystem)] TypedActor Started
[INFO] [03/21/2016 22:41:51.124] [main] [Example12_10$(akka://TypedActorSys
tem)] fSquare=100
[INFO] [03/21/2016 22:41:51.131] [TypedActorSystem-akka.actor.default-dispa
tcher-5] [$Proxy1(akka://TypedActorSystem)] TypedActor Stopped
[INFO] [03/21/2016 22:41:51.131] [TypedActorSystem-akka.actor.default-dispa
tcher-3] [$Proxy0(akka://TypedActorSystem)] TypedActor Stopped
```

代码中类 SquarerImpl 混入了 PreStart 和 PostStop 两个 trait：class SquarerImpl(val name: String) extends Squarer with PostStop with PreStart，这样的话在创建 TypedActor 之前和停止 TypedActor 后能够进行相应的操作，本例中主要是为监视 TypedActor 的创建和停止过程。代码 TypedActor(system).stop(mySquarer)通过 stop 方法停止 TypedActor，而 TypedActor(system).poisonPill(otherSquarer)通过调用 poisonPill 方法停止运行 TypedActor。

12.4 Dispatcher

12.4.1 常用 Dispatcher

Akka 中的 Dispatcher 主要用于接收消息并对消息进行分发，使用 Dispatcher 能够控制程序的执行。每个 ActorSystem 中都有对应的 Dispatcher，如果在创建 ActorSystem 时不指定 Dispatcher，则使用默认的 Dispatcher。默认的 Dispatcher 配置信息如下:

```
Akka-Default-Dsipatcher-Example{
defaultDispatcher {
        //指定 Dispatcher 的类型，type = Dispatcher 表示使用的是默认的 Dispatcher
        type = Dispatcher
        //指定 ExecutionService 的类型
        executor = "fork-join-executor"
        // fork join pool 配置信息
        fork-join-executor {
```

```
            //与并行度乘积因子一起,控制最小线程数
            parallelism-min = 2
            //并行度乘积因子,
            parallelism-factor = 2.0
            //与并行度乘积因子一起,控制最大线程数
            parallelism-max = 6
        }
    }
}
```

除了默认的 Dispatcher 之外,还有 3 种类型的 Dispatcher,它们分别是 BalancingDispatcher、PinnedDispatcher 和 CallingThreadDispatcher。具体说明如表 12-1 所示。

表 12-1 Dispatcher 的类型

Dispatcher 类型	说明
Dispatcher	描述:Akka 中默认的 Dispatcher,不指定 Dispatcher 时使用 共享度:无限制,线程池由若干个 Actor 共享 Mailbox:每个 Actor 都有自己对应的邮箱(Mailbox) 支持的 ExecutorService 类型:fork-join-executor、thread-pool-executor 及自定义扩展自 akka.dispatcher.ExecutorServiceConfigurator 的类 使用场景:非阻塞式的应用
PinnedDispatcher	描述: 为每个 Actor 分配唯一的线程,也就是每个 Actor 对应一个只有一个线程的线程池 共享度:线程专用,不共享 Mailbox:每个 Actor 都有自己对应的邮箱(Mailbox) 支持的 ExecutorService 类型:thread-pool-executor 使用场景:IO 操作、长运行任务
BalancingDispatcher	描述: 将运行任务从运行繁忙的 Actor 分发给空闲的 Actor 共享度:由同类型的 Actor 共享 Mailbox:多个 Actor 共用一个 Mailbox 支持的 ExecutorService 类型:fork-join-executor、thread-pool-executor 及自定义扩展自 akka.dispatcher.ExecutorServiceConfigurator 的类 使用场景:协同任务
CallingThreadDispatcher	描述: 在当前线程上运行任务 共享度:无限制,由若干个 Actor 共享 Mailbox:每个 Actor 都有自己对应的邮箱(Mailbox) 支持的 ExecutorService 类型:当前运行的线程 使用场景:测试

默认 Dispatcher 的使用示例如下:

```
/*
 * Dispatcher:默认 Dispatcher
 */
```

```scala
object Example12_11 extends App {
  import akka.actor.ActorSystem
  import com.typesafe.config.ConfigFactory
  import akka.actor.Props

  class StringActor extends Actor {
    val log = Logging(context.system, this)

    def receive = {
      case s:String   {
        log.info("received message:\n"+s)
      };
      case _      log.info("received unknown message")
    }

    override def postStop(): Unit = {
      log.info("postStop in StringActor")
    }
  }

  //从 application.conf 配置文件中加载 dispatcher 配置信息
  val _system
  = ActorSystem.create("DsipatcherSystem",ConfigFactory.load().getConfig("Akka-Default-Dsipatcher-Example"))

  //创建 Actor 时通过 withDispatcher 方法指定自定义的 Dispatcher
  val stringActor = _system.actorOf(Props[StringActor].withDispatcher("defaultDispatcher"),name="StringActor")

  stringActor!"Test"

  _system.shutdown()
}
```

代码运行结果如下：

[INFO] [03/22/2016 16:55:27.557] [DsipatcherSystem-defaultDispatcher-6] [akka://DsipatcherSystem/user/StringActor] received message:
Test
[INFO] [03/22/2016 16:55:27.608] [DsipatcherSystem-defaultDispatcher-6] [akka://DsipatcherSystem/user/StringActor] postStop in StringActor

代码

ActorSystem.create("DsipatcherSystem",ConfigFactory.load().getConfig("Akka

```
-Default-Dsipatcher-Example"))
```

通过配置文件创建 ActorSystem，ConfigFactory.load().getConfig("Akka-Default-Dsipatcher-Example")) 将 application.conf 中的配置信息加载到内存中，在创建 Actor 时通过 Props[StringActor].withDispatcher("defaultDispatcher")指定定义的 Dispatcher。

12.4.2 ExecutionService

Akka 中的 Dispatcher 基于 Java 语言中的 Executor（java.util.concurrent），Executor 基于生产者-消费者模型，从而达到将任务的提交和执行进行解耦的目的。Akka 中一个典型的 ForkJoinPool 配置如下所示：

```
executor = "fork-join-executor"
//线程池配置
fork-join-executor {
    parallelism-min = 2
    parallelism-factor = 2.0
    parallelism-max = 6
}
```

Dispatcher 开启的最少线程数为 parallelism-min*CPU 核数* parallelism-factor，开启的最大线程数为 parallelism-max *CPU 核数* parallelism-factor。一个典型的 ThreadPoolExecutor 定义如下：

```
executor = "thread-pool-executor"
//线程池配置
thread-pool-executor {
        // 与并行度乘积因子一起，控制最小线程数
        core-pool-size-min = 2
        //并行度乘积因子
        core-pool-size-factor = 2.0
        // 与并行度乘积因子一起，控制最大线程数
    core-pool-size-max = 10
}
```

Dispatcher 开启的线程数与 ForkJoinPool 中的 Dispatcher 开启的线程数类似。ForkJoinPool 和 ThreadPoolExecutor 配置时还有几个重要的参数可以配置，如下所示：

```
my-thread-pool-dispatcher {
# 默认 Dispatcher
type = Dispatcher
# ExecutionService 类型
executor = "thread-pool-executor"
# 线程池配置
thread-pool-executor {
```

```
    core-pool-size-min = 2
    core-pool-size-factor = 2.0
    core-pool-size-max = 10
}
# 吞吐量：线程处理下一个 Actor 消息之前，每个 Actor 上处理的最大消息数
throughput = 100
# （可选参数），指定 mailbox 的容量，负数或零表示邮箱是无边界的（Unbounded）；正数表示指定
大小的邮箱，邮箱是有边界的（Bounded）
mailbox-capacity = -1
# （可选参数），指定邮箱的类型
mailbox-type =""
}
```

有以下几种类型的邮箱：

- UnboundedMailbox。使用 java.util.concurrent.ConcurrentLinkedQueue 实现，是一种非阻塞式的、无边界的 Mailbox。
- BoundedMailbox。使用 java.util.concurrent.LinkedBlockingQueue 实现，是一种阻塞式的、有边界的 Mailbox。
- UnboundedPriorityMailbox。使用 java.util.concurrent.PriorityBlockingQueue 实现，是一种阻塞式的、无边界 Mailebox。
- BoundedPriorityMailbox。使用 akka.util.BoundedBlockingQueu 进行实现，它是对 java.util.PriorityBlockingQueue 的封装，是一种阻塞式的、有边界的邮箱。

下面的代码给出的是 UnboundedPriorityMailbox 的使用示例。

```
/*
 * UnboundedPriorityMailbox 使用
 */
object Example12_12 extends App {
  import akka.dispatch.PriorityGenerator
  import akka.dispatch.UnboundedPriorityMailbox
  import com.typesafe.config.Config

  //自定义 Mailbox,扩展自 UnboundedPriorityMailbox
  class MyPrioMailbox(settings: ActorSystem.Settings, config: Config)
    extends UnboundedPriorityMailbox(
      // 创建 PriorityGenerator，值越低表示优先级越高
      PriorityGenerator {
        // 'highpriority 为符号信息，首先处理（高优先级）
        case 'highpriority =>0
        // 'lowpriority 为符号信息，最后处理（低优先级）
        case 'lowpriority =>2
        // PoisonPill 停止 Actor 运行
```

```
            case PoisonPill=>3
            // 默认优先级，值介于高优先级和低优先级之间
            case otherwise => 1
        })


        //从application.conf 配置文件中加载dispatcher配置信息
        val _system = ActorSystem.create("DsipatcherSystem",ConfigFactory.load().getConfig("MyDispatcherExample"))

        // We create a new Actor that just prints out what it processes
        val a = _system.actorOf(
          Props(new Actor {
            val log: LoggingAdapter = Logging(context.system, this)
            self ! 'lowpriority
            self ! 'lowpriority
            self ! 'highpriority
            self ! 'pigdog
            self ! 'pigdog2
            self ! 'pigdog3
            self ! 'highpriority
            self ! PoisonPill
            def receive = {
              case x => log.info(x.toString)
            }
          }).withDispatcher("balancingDispatcher"),name="UnboundedPriorityMailboxActor")

        _system.shutdown()
      }
```

代码运行结果如下：

    [INFO] [03/22/2016 21:12:14.874] [DsipatcherSystem-balancingDispatcher-6] [akka://DsipatcherSystem/user/UnboundedPriorityMailboxActor] 'highpriority
    [INFO] [03/22/2016 21:12:14.879] [DsipatcherSystem-balancingDispatcher-6] [akka://DsipatcherSystem/user/UnboundedPriorityMailboxActor] 'highpriority
    [INFO] [03/22/2016 21:12:14.879] [DsipatcherSystem-balancingDispatcher-6] [akka://DsipatcherSystem/user/UnboundedPriorityMailboxActor] 'pigdog
    [INFO] [03/22/2016 21:12:14.880] [DsipatcherSystem-balancingDispatcher-6] [akka://DsipatcherSystem/user/UnboundedPriorityMailboxActor] 'pigdog2
    [INFO] [03/22/2016 21:12:14.880] [DsipatcherSystem-balancingDispatcher-6] [akka://DsipatcherSystem/user/UnboundedPriorityMailboxActor] 'pigdog3
    [INFO] [03/22/2016 21:12:14.880] [DsipatcherSystem-balancingDispatcher-7]

```
[akka://DsipatcherSystem/user/UnboundedPriorityMailboxActor] 'lowpriority
```

代码中创建 ActorSystem 实例时使用的是 ConfigFactory 加载配置文件，配置文件中的 MyDispatcherExample 的内容如下：

```
MyDispatcherExample{
balancingDispatcher {
      type = BalancingDispatcher
      executor = "thread-pool-executor"
      thread-pool-executor {
          core-pool-size-min = 1
          core-pool-size-factor = 2.0
          core-pool-size-max = 2
      }
      throughput = 5
      mailbox-type = "cn.scala.chapter12.Example12_12$MyPrioMailbox"
   }
}
```

在创建 Actor 时使用 withDispatcher("balancingDispatcher")方法指定其 Dispatcher 为 BalancingDispatcher，通过观察发送符号消息的顺序和日志输出的消息顺序可以看到，消息的处理是按照优先级进行的，高优先级的消息先被处理，而优先级低的消息后处理。

12.5 Router

Dispatcher 用于提高程序处理消息的吞吐量，而本节介绍的 Router 用于将输入的消息进行路由转发，交由外围的 Actor（也称 Routee）处理，提高了 Actor 并行处理消息的能力。Akka 自身提供了以下几种类型的 Router：

- akka.routing.RoundRobinRouter 将进入的消息循环路由到所有的 Router 上。
- akka.routing.RandomRouter 随机选取一个 Router 并路由消息到该 Router 上。
- akka.routing.SmallestMailboxRouter 选择邮箱中消息最少的 Router 并路由消息到该 Router 上。
- akka.routing.BroadcastRouter 将同样的消息路由给所有的 Router。
- akka.routing.ScatterGatherFirstCompletedRouter 将消息路由给所有的 Router，返回结果存为 Future，任意一 Router 返回结果时将该结果立即返回给调用者。
- akka.routing.ConsistentHashingRouter 使用一致性哈希算法选择 Router 并将消息路由给该 Router。

下面的代码演示的是 akka.routing.RandomRouter 的使用。

```
/*
 * RandomRouter
```

```scala
    */
    object Example12_13 extends App {
      import akka.actor.ActorSystem
      import akka.actor.Props
      import akka.routing.RandomRouter

      class IntActor extends Actor {
        val log = Logging(context.system, this)

        def receive = {
          case s: Int   {
            log.info("received message:\n" + s)
          }
          case _   log.info("received unknown message")
        }
      }
      val _system = ActorSystem("RandomRouterExample")
      // withRouter 方法指定 Router 类型,本例中为 RandomRouter (5),5 表示 Routee 对象数
      val randomRouter = _system.actorOf(Props[IntActor].withRouter(RandomRouter(5)), name = "IntActor")
      1 to 10 foreach {
        i => randomRouter ! i
      }
      _system.shutdown()

    }
```

程序运行结果如下:

```
[INFO] [03/22/2016 22:50:49.191] [RandomRouterExample-akka.actor.default-dispatcher-12] [akka://RandomRouterExample/user/IntActor/$c] received message:1
[INFO] [03/22/2016 22:50:49.200] [RandomRouterExample-akka.actor.default-dispatcher-12] [akka://RandomRouterExample/user/IntActor/$c] received message:4
[INFO] [03/22/2016 22:50:49.200] [RandomRouterExample-akka.actor.default-dispatcher-12] [akka://RandomRouterExample/user/IntActor/$c] received message:7
[INFO] [03/22/2016 22:50:49.200] [RandomRouterExample-akka.actor.default-dispatcher-12] [akka://RandomRouterExample/user/IntActor/$a] received message:2
[INFO] [03/22/2016 22:50:49.201] [RandomRouterExample-akka.actor.default-dispatcher-12] [akka://RandomRouterExample/user/IntActor/$b] received message:3
[INFO] [03/22/2016 22:50:49.201] [RandomRouterExample-akka.actor.default-dispatcher-12] [akka://RandomRouterExample/user/IntActor/$b] received message:6
[INFO] [03/22/2016 22:50:49.201] [RandomRouterExample-akka.actor.default-dispatcher-12] [akka://RandomRouterExample/user/IntActor/$b] received message:8
[INFO] [03/22/2016 22:50:49.201] [RandomRouterExample-akka.actor.default-di
```

```
spatcher-12] [akka://RandomRouterExample/user/IntActor/$b] received message:10
    [INFO] [03/22/2016 22:50:49.202] [RandomRouterExample-akka.actor.default-di
spatcher-12] [akka://RandomRouterExample/user/IntActor/$e] received message:5
    [INFO] [03/22/2016 22:50:49.202] [RandomRouterExample-akka.actor.default-di
spatcher-12] [akka://RandomRouterExample/user/IntActor/$e] received message:9
```

代码 val randomRouter = _system.actorOf(Props[IntActor].withRouter(RandomRouter(5)), name = "IntActor")通过 withRouter 方法指定使用的 RandomRouter，可以看到，IntActor（Router）会产生 5 个 Routee，本例中对应的是 [akka://RandomRouterExample/user/IntActor/$a]~[akka://RandomRouterExample/user/IntActor/$e]共 5 个 Actor 实例。我们知道，RandomRouter会随机选取一个 Router，然后将消息路由给该 Router 处理，因此多次运行 RandomRouter 会发现其每次运行结果不同且可能 Router 个数会小于 5。

12.6 容 错

12.6.1 Actor 的 4 种容错机制

Akka 提供"Let it crash"容错机制，当 Actor 抛出异常时，根据设定的容错机制进行相应的操作，目前 Akka 支持 4 种容错机制，分别是重启（Restart）、恢复（Resume）、停止（Terminate）及上报（Escalate）。Restart 指的是 Supervisor 将出错的 Actor kill 掉，然后重新创建一个新的Actor；Resume 指的是保留出错的 Actor 状态并恢复到出错时的状态；Terminate 指的是永久地停止当前的 Actor 运行；Escalate 指的是将错误上报，主要针对的是当前 Supervisor 无法处理的异常情况。

12.6.2 Supervison

Akka 支持两种监控策略（Supervision Strategy）：akka.actor.OneForOneStrategy 和 akka.actor.AllForOneStrategy。OneForOneStrategy 指某 Supervisor 的各子 Actor 在出错时，只对出错的子 Actor 根据指定的策略进行处理，而不影响其他兄弟子 Actor；AllForOneStrategy 指某 Supervisor 的各子 Actor 在出错时，所有的子 Actor 都会根据预定的策略进行处理。一个典型的 OneForOneStrategy 的定义如下所示：

```
import akka.actor.OneForOneStrategy
import akka.actor.SupervisorStrategy._
import scala.concurrent.duration._
override val supervisorStrategy =OneForOneStrategy(maxNrOfRetries = 10, wit
hinTimeRange = 1 minute) {
        case _: ArithmeticException =>Resume
        case _: NullPointerException =>Restart
        case _: IllegalArgumentException  =>Stop
```

```
        case _: Exception => Escalate
}
```

代码 OneForOneStrategy(maxNrOfRetries = 10, withinTimeRange = 1 minute) 中的 maxNrOfRetries 参数表示最大尝试对应策略的次数，withinTimeRange 参数表示重启动的有效时间范围，以下参数

```
{
    case _: ArithmeticException => Resume
    case _: NullPointerException => Restart
    case _: IllegalArgumentException => Stop
    case _: Exception => Escalate
}
```

为指定的具体策略，它是一种 SupervisorStrategy.Decider 类型，被定义为 type Decider = PartialFunction[Throwable, Directive]，当出错类型为 ArithmeticException 时进行 Resume（恢复）操作，出错类型为 NullPointerException 时进行 Restart（重启）操作，出错类型为 IllegalArgumentException 时进行 Stop（停止）操作，其他出错类型则进行 Escalate（上报）操作。AllForOneStrategy 策略的定义与 OneForOneStrategy 策略类似，例如：

```
override val supervisorStrategy = AllForOneStrategy(maxNrOfRetries = 10, withinTimeRange = 10 seconds) {
    case _: ArithmeticException => Resume
    case _: NullPointerException => Restart
    case _: IllegalArgumentException => Stop
    case _: Exception => Escalate
}
```

不同的地方仅在于 OneForOneStrategy 策略只会作用于出错的 Actor，而 AllForOneStrategy 会作用于出错的 Actor 以及所有处于同一个 Supervisor 下的兄弟 Actor。

OneForOneStrategy 策略的具体使用代码如下：

```scala
/*
 * 容错：Supervisor 策略，One-For-One strategy
 */
object Example13_14 extends App {
  import akka.actor.actorRef2Scala
  import akka.actor.Actor
  import akka.actor.ActorLogging
  import akka.actor.Props
  import scala.concurrent.duration._
  import akka.pattern.ask

  case class NormalMessage()
  class ChildActor extends Actor with ActorLogging {
```

```scala
    var state: Int = 0

    override def preStart() {
      log.info("启动 ChildActor, 其 hashcode 为"+this.hashCode())
    }
    override def postStop() {
      log.info("停止 ChildActor, 其 hashcode 为"+this.hashCode())
    }
    def receive: Receive = {
      case value: Int =>
        if (value <= 0)
          throw new ArithmeticException("数字小于等于0")
        else
          state = value
      case result: NormalMessage =>
        sender ! state
      case ex: NullPointerException =>
        throw new NullPointerException("空指针")
      case _ =>
        throw new IllegalArgumentException("非法参数")
    }
  }

  class SupervisorActor extends Actor with ActorLogging {
    import akka.actor.OneForOneStrategy
    import akka.actor.SupervisorStrategy._
    val childActor = context.actorOf(Props[ChildActor], name = "ChildActor")

    override val supervisorStrategy = OneForOneStrategy(maxNrOfRetries = 10, withinTimeRange = 10 seconds) {
      //异常类型为 ArithmeticException 时, 采用 Resume 机制
      case _: ArithmeticException => Resume
      //异常类型为 NullPointerException 时, 采用 Resume 机制
      case _: NullPointerException => Restart
      //异常类型为 IllegalArgumentException 时, 采用 Stop 机制
      case _: IllegalArgumentException => Stop
      //其他异常机制, 采用 Escalate 机制
      case _: Exception => Escalate
    }

    def receive = {
      case msg: NormalMessage =>
```

```scala
            childActor.tell(msg, sender)
          case msg: Object =>
            childActor ! msg

      }
    }

    val system = ActorSystem("FaultToleranceSystem")
    val log = system.log

    val supervisor = system.actorOf(Props[SupervisorActor], name = "SupervisorActor")

    //正数，消息正常处理
    supervisor ! 5
    implicit val timeout = Timeout(5 seconds)
    var future = (supervisor ? new NormalMessage).mapTo[Int]
    var resultMsg = Await.result(future, timeout.duration)
    log.info("结果:"+resultMsg)

    //负数，Actor 会抛出异常，Superrvisor 使用 Resume 处理机制
    supervisor ! -5
    future = (supervisor ? new NormalMessage).mapTo[Int]
    resultMsg = Await.result(future, timeout.duration)
    log.info("结果:"+resultMsg)

    //空指针消息，Actor 会抛出异常，Superrvisor 使用 restart 处理机制
    supervisor ! new NullPointerException
    future = (supervisor ? new NormalMessage).mapTo[Int]
    resultMsg = Await.result(future, timeout.duration)
    log.info("结果:"+resultMsg)

    //String 类型参数为非法参数，Actor 会抛出异常，Superrvisor 使用 stop 处理机制
    supervisor ? "字符串"

    system.shutdown
  }
```

程序运行结果如下：

[INFO] [03/27/2016 23:18:26.074] [FaultToleranceSystem-akka.actor.default-dispatcher-2] [akka://FaultToleranceSystem/user/SupervisorActor/ChildActor] 启动 ChildActor，其 hashcode 为 30120695
[INFO] [03/27/2016 23:18:26.093] [main] [ActorSystem(FaultToleranceSystem)]

结果:5
    [ERROR] [03/27/2016 23:18:26.093] [FaultToleranceSystem-akka.actor.default-dispatcher-2] [akka://FaultToleranceSystem/user/SupervisorActor/ChildActor] 数字小于等于0
    java.lang.ArithmeticException: 数字小于等于0
        at cn.scala.chapter13.Example13_14$ChildActor$$anonfun$receive$19.applyOrElse(Example13_1.scala:717)
        at akka.actor.ActorCell.receiveMessage(ActorCell.scala:425)
        at akka.actor.ActorCell.invoke(ActorCell.scala:386)
        at akka.dispatch.Mailbox.processMailbox(Mailbox.scala:230)
        at akka.dispatch.Mailbox.run(Mailbox.scala:212)
        at akka.dispatch.ForkJoinExecutorConfigurator$MailboxExecutionTask.exec(AbstractDispatcher.scala:506)
        at scala.concurrent.forkjoin.ForkJoinTask.doExec(ForkJoinTask.java:260)
        at scala.concurrent.forkjoin.ForkJoinPool$WorkQueue.runTask(ForkJoinPool.java:1339)
        at scala.concurrent.forkjoin.ForkJoinPool.runWorker(ForkJoinPool.java:1979)
        at scala.concurrent.forkjoin.ForkJoinWorkerThread.run(ForkJoinWorkerThread.java:107)

    [INFO] [03/27/2016 23:18:26.116] [main] [ActorSystem(FaultToleranceSystem)]
结果:5
    [ERROR] [03/27/2016 23:18:26.117] [FaultToleranceSystem-akka.actor.default-dispatcher-5] [akka://FaultToleranceSystem/user/SupervisorActor/ChildActor] 空指针
    java.lang.NullPointerException: 空指针
        at cn.scala.chapter13.Example13_14$ChildActor$$anonfun$receive$19.applyOrElse(Example13_1.scala:723)
        at akka.actor.ActorCell.receiveMessage(ActorCell.scala:425)
        at akka.actor.ActorCell.invoke(ActorCell.scala:386)
        at akka.dispatch.Mailbox.processMailbox(Mailbox.scala:230)
        at akka.dispatch.Mailbox.run(Mailbox.scala:212)
        at akka.dispatch.ForkJoinExecutorConfigurator$MailboxExecutionTask.exec(AbstractDispatcher.scala:506)
        at scala.concurrent.forkjoin.ForkJoinTask.doExec(ForkJoinTask.java:260)
        at scala.concurrent.forkjoin.ForkJoinPool$WorkQueue.runTask(ForkJoinPool.java:1339)
        at scala.concurrent.forkjoin.ForkJoinPool.runWorker(ForkJoinPool.java:1979)
        at scala.concurrent.forkjoin.ForkJoinWorkerThread.run(ForkJoinWorkerThread.java:107)

```
[INFO] [03/27/2016 23:18:26.118] [FaultToleranceSystem-akka.actor.default-d
ispatcher-2] [akka://FaultToleranceSystem/user/SupervisorActor/ChildActor] 停止
ChildActor，其 hashcode 为 30120695
[INFO] [03/27/2016 23:18:26.120] [FaultToleranceSystem-akka.actor.default-d
ispatcher-2] [akka://FaultToleranceSystem/user/SupervisorActor/ChildActor] 启动
ChildActor，其 hashcode 为 13650807
[INFO] [03/27/2016 23:18:26.120] [main] [ActorSystem(FaultToleranceSystem)]
结果：0
[ERROR] [03/27/2016 23:18:26.124] [FaultToleranceSystem-akka.actor.default-
dispatcher-5] [akka://FaultToleranceSystem/user/SupervisorActor/ChildActor] 非
法参数
java.lang.IllegalArgumentException: 非法参数
    at cn.scala.chapter13.Example13_14$ChildActor$$anonfun$receive$19.apply
OrElse(Example13_1.scala:725)
    at akka.actor.ActorCell.receiveMessage(ActorCell.scala:425)
    at akka.actor.ActorCell.invoke(ActorCell.scala:386)
    at akka.dispatch.Mailbox.processMailbox(Mailbox.scala:230)
    at akka.dispatch.Mailbox.run(Mailbox.scala:212)
    at akka.dispatch.ForkJoinExecutorConfigurator$MailboxExecutionTask.exec
(AbstractDispatcher.scala:506)
    at scala.concurrent.forkjoin.ForkJoinTask.doExec(ForkJoinTask.java:260)
    at scala.concurrent.forkjoin.ForkJoinPool$WorkQueue.runTask(ForkJoinPoo
l.java:1339)
    at scala.concurrent.forkjoin.ForkJoinPool.runWorker(ForkJoinPool.java:1
979)
    at scala.concurrent.forkjoin.ForkJoinWorkerThread.run(ForkJoinWorkerThr
ead.java:107)

[INFO] [03/27/2016 23:18:26.128] [FaultToleranceSystem-akka.actor.default-d
ispatcher-5] [akka://FaultToleranceSystem/user/SupervisorActor/ChildActor] 停止
ChildActor，其 hashcode 为 13650807
```

当向 SupervisorActor 发送正整数 5 时，childActor 能够正常处理；当向 SupervisorActor 发送 -5 时，childActor 会出现 ArithmeticException 异常。由于采用的是 Resume 机制，因此 SupervisorActor 会尝试进行恢复，恢复到出错之前的状态，输出信息为：

```
[ERROR] [03/27/2016 23:18:26.093] [FaultToleranceSystem-akka.actor.default-
dispatcher-2] [akka://FaultToleranceSystem/user/SupervisorActor/ChildActor] 数
字小于等于 0
java.lang.ArithmeticException: 数字小于等于 0
    at cn.scala.chapter13.Example13_14$ChildActor$$anonfun$receive$19.apply
OrElse(Example13_1.scala:717)
    //部分代码略
```

```
[INFO] [03/27/2016 23:18:26.116] [main] [ActorSystem(FaultToleranceSystem)]
结果:5
```

当向 SupervisorActor 发送 new NullPointerException 对象时，childActor 会出现 NullPointerException 异常，此时采用的便是 Restart 机制，将当前的 Actor 杀死，然后新创建一个 ChildActor，因此对应的输出结果如下：

```
[ERROR] [03/27/2016 23:18:26.117] [FaultToleranceSystem-akka.actor.default-dispatcher-5] [akka://FaultToleranceSystem/user/SupervisorActor/ChildActor] 空指针
java.lang.NullPointerException: 空指针
    at cn.scala.chapter13.Example13_14$ChildActor$$anonfun$receive$19.applyOrElse(Example13_1.scala:723)
    //部分代码略

[INFO] [03/27/2016 23:18:26.118] [FaultToleranceSystem-akka.actor.default-dispatcher-2] [akka://FaultToleranceSystem/user/SupervisorActor/ChildActor] 停止 ChildActor，其 hashcode 为 30120695
[INFO] [03/27/2016 23:18:26.120] [FaultToleranceSystem-akka.actor.default-dispatcher-2] [akka://FaultToleranceSystem/user/SupervisorActor/ChildActor] 启动 ChildActor，其 hashcode 为 13650807
[INFO] [03/27/2016 23:18:26.120] [main] [ActorSystem(FaultToleranceSystem)]
结果:0
```

当向 SupervisorActor 发送字符串消息时：supervisor ? "字符串"，ChildActor 会报 IllegalArgumentException 非法参数异常的错误，此时采用的是 Stop 机制，因此其输出信息为

```
[ERROR] [03/27/2016 23:18:26.124] [FaultToleranceSystem-akka.actor.default-dispatcher-5] [akka://FaultToleranceSystem/user/SupervisorActor/ChildActor] 非法参数
java.lang.IllegalArgumentException: 非法参数
    //部分代码略
[INFO] [03/27/2016 23:18:26.128] [FaultToleranceSystem-akka.actor.default-dispatcher-5] [akka://FaultToleranceSystem/user/SupervisorActor/ChildActor] 停止 ChildActor，其 hashcode 为 13650807
```

至于 AllForOneStrategy 策略的使用，与 OneForOneStrategy 类似，只不过 AllForOneStrategy 影响面更广，不仅影响到出错的子 Actor，还影响到它所有的兄弟 Actor，感兴趣的读者可以自行实现观察程序的行为。

小　结

本章对 Java 语言中并发应用程序编写的常见问题进行了分析，然后介绍了 Scala 语言并发编程模型中的 Actor 及 Akka 并发框架，对 Actor、Typed Actor、Dispatcher、Router 及容错等并发编程基础内容进行了介绍，让读者对 Scala 并发编程的特点有了基本的认识。

第13章 Scala 与 Java 的互操作

　　Scala 语言与 Java 语言有着良好的互操作性，Scala 语言几乎完全兼容 Java 语言，同时又有自身独有的特点，这导致它们在互操作时需要进行部分特殊的处理。本章将详细介绍 Scala 与 Java 间互操作时需要特别注意的几个方面的内容。

13.1 Java 与 Scala 集合互操作

无论是何种编程语言，集合（collections）都是最常用的数据结构。在第 4 章中我们重点介绍了 Scala 语言中的集合，如果读者熟悉 Java 语言的话，可以明显地感觉到 Java 语言与 Scala 语言集合间的两个主要区别：首先，Scala 集合有两种，分别是 immutable 和 mutable 集合，Scala 语言推崇使用 immutable 集合；其次，Scala 集合中拥有大量 Java 语言中没有的方法。本小节将主要介绍 Java 与 Scala 集合间的互操作问题。

13.1.1 Java 调用 Scala 集合

下面的代码演示的是 Java 语言中如何使用 Scala 集合，重点演示如何将 Scala 集合转换成 Java 集合，进而通过 Java 语言中的 foreach 语法对集合进行遍历。

```java
package cn.scala.chapter13;
import scala.Function1;
import scala.Tuple2;
import scala.collection.JavaConversions;
import scala.collection.mutable.HashMap;
import scala.collection.mutable.Map;
/**
 * 在 Java 代码中调用 Scala 集合
 */
public class ScalaCollectionInJava {
    public static void main(String[] args) {
       Map<String,String> scalaBigDataTools=new HashMap<>();
        scalaBigDataTools.put("Hadoop","the most popular big data processing tools");
        scalaBigDataTools.put("Hive","the most popular interactive query tools");

        //scala.collection.mutable.Map 不能使用 Java 语法中的 for each 语句对集合中的元素进行遍历
        /* for (String key: scalaBigDataTools.keySet()) {
            System.out.println(scalaBigDataTools.get(key));
        }*/

        //使用 scala 集合中提供的 foreach 方法，但需要自己实现相应的函数，处理方式较为复杂
        /* scalaBigDataTools.foreach(new Function1<Tuple2<String, String>, Object>(){
```

```
        });*/

        //将 scala.collection.mutable.Map 转换成 java.util.Map
        java.util.Map<String,String> javaBigDataTools =JavaConversions.asJavaMap(scalaBigDataTools);
        //此时便可以使用 for each 语句对集合进行遍历
        for (String key: javaBigDataTools.keySet()) {
            System.out.println(javaBigDataTools.get(key));
        }
    }
}
```

代码中 Map<String,String> scalaBigDataTools=new HashMap<>()使用的是 Scala 集合 scala.collection.mutable.HashMap，使用时的主要问题是如何进行集合的遍历，它不能直接通过 Java 中的 foreach 语法对集合进行遍历，例如：

```
for (String key: scalaBigDataTools.keySet()) {
        System.out.println(scalaBigDataTools.get(key));
}
```

而调用 scala.collection.mutable.HashMap 的方法又需要自己实现对应的函数，使用起来较为复杂。解决方法便是引入 scala.collection.JavaConversions 类，然后调用 JavaConversions.asJavaMap 方法将 scala.collection.mutable.HashMap 转换成 Java 语言的 java.util.Map，再通过 for (String key: javaBigDataTools.keySet())对集合的元素进行遍历。

13.1.2　Scala 调用 Java 集合

Scala 语言中可以直接使用 Java 语言中的现有类，只要符合 Scala 的基本语法即可。下列代码给出的是在 Scala 中如何使用 Java 集合，重点仍然是如何对集合的元素进行遍历。

```
package cn.scala.chapter13
import java.util.ArrayList;
object JavaCollectionInScala extends App{
  def getList= {
    val list = new ArrayList[String]()
    list.add("Hadoop")
    list.add("Hive")
    list
  }
  val list=getList
  //因为 list 是 java.util.ArrayList 类型
  //不能调用 scala 集合的 foreach 方法
  //list.foreach(println)
```

从上述代码中可以看到,Scala 在使用 Java 集合中的方法时,只要符合 Scala 语法即可,但它不能直接使用 Scala 集合中的 foreach 方法,如果想使用 Scala 语言中提供的集合遍历语法,也需要对集合进行转换,转换方式同样也是通过引入 import scala.collection.JavaConversions._,只不过 Scala 语言可以通过隐式转换自动地将 Java 集合转换成 Scala 集合,具体见下面的代码。

```
package cn.scala.chapter13
import java.util.ArrayList;
object JavaCollectionInScala extends App{
  def getList= {
    val list = new ArrayList[String]()
    list.add("Hadoop")
    list.add("Hive")
    list
  }
  val list=getList
  //引入 scala.collection.JavaConversions._ 后
  //便可以使用 Scala 集合提供的一系列方法
  import scala.collection.JavaConversions._
  list.foreach(println)
  val list2=list.map(x=>x*2)
  println(list2)
}
```

代码运行结果如下:

```
Hadoop
Hive
ArrayBuffer(HadoopHadoop, HiveHive)
```

13.1.3　Scala 与 Java 集合间相互转换分析

通过 Scala 中调用 Java 集合和 Java 中调用 Scala 集合这两个例子可以看出,scala.collection.JavaConversions 类在其中起了十分重要的作用,在 Scala 中使用时可以借助隐式转换自动进行,而在 Java 中由于不支持隐式转换需要手动调用将 Scala 集合转换成 Java 集合。在转换过程中,部分转换是双向的,而其他部分转换是单向的,所谓转换具有双向性是指将 Scala 集合转换成 Java 集合,再将其又转换成 Scala 集合,得到的集合与原来的 Scala 集合是同一个,而单向转换则没有这一特点。

下面的代码给出的是双向转换的示例:

```
scala> :paste
// Entering paste mode (ctrl-D to finish)
import scala.collection.JavaConversions._
//创建 Scala 集合
```

```
val s1 = new scala.collection.mutable.ListBuffer[Int]
//通过隐式转换为 Java 集合
val jl : java.util.List[Int] = s1
//再通过隐式转换为 Scala 集合
val sl2 : scala.collection.mutable.Buffer[Int] = jl
// Exiting paste mode, now interpreting.
import scala.collection.JavaConversions._
s1: scala.collection.mutable.ListBuffer[Int] = ListBuffer()
jl: java.util.List[Int] = []
sl2: scala.collection.mutable.Buffer[Int] = ListBuffer()

//双向转换后的 sl2 与原来的 Scala 集合 s1 为同一个对象
scala> s1 eq sl2
res1: Boolean = true
```

JavaConversions 类支持的双向转换包括:

scala.collection.Iterable	<=>	java.lang.Iterable
scala.collection.Iterable	<=>	java.util.Collection
scala.collection.Iterator	<=>	java.util.{ Iterator, Enumeration }
scala.collection.mutable.Buffer	<=>	java.util.List
scala.collection.mutable.Set	<=>	java.util.Set
scala.collection.mutable.Map	<=>	java.util.{ Map, Dictionary }
scala.collection.mutable.ConcurrentMap	<=>	java.util.concurrent.ConcurrentMap
scala.collection.concurrent.Map	<=>	java.util.concurrent.ConcurrentMap

JavaConversions 类支持的单向转换包括:

scala.collection.Seq	=>	java.util.List
scala.collection.mutable.Seq	=>	java.util.List
scala.collection.Set	=>	java.util.Set
scala.collection.Map	=>	java.util.Map
java.util.Properties	=>	scala.collection.mutable.Map[String, String]

13.2 Scala 与 Java 泛型互操作

熟悉 Java 语言的读者可能会发现:Scala 语言与 Java 语言的泛型更多地只是语法上的区别,如 Java 语言中的泛型语法为类名<泛型>,例如 List<String>,Scala 语言中的泛型语法为类名[泛型],例如 List[String]。本小节将重点介绍 Scala 与 Java 间的泛型是如何进行互操作的。

13.2.1 Scala 中使用 Java 泛型

下面的代码演示的是 Scala 语言中如何使用 Java 中的泛型类。

```
package cn.scala.chapter13
import java.util._

case class Person(val name:String,val age:Int)

//在 Java 中 Comparator 是这么用的：Comparator<Person>
//而在 Scala 中是这么用的：Comparator[Person]
class PersonComparator extends Comparator[Person]{
  override def compare(o1: Person, o2: Person): Int = if(o1.age>o2.age) 1 else -1
}
/**
 * Java 泛型在 Scala 中的使用示例
 */
object JavaGenericInScala extends App{
  val p1=Person("摇摆少年梦",27)
  val p2=Person("李四",29)
  val personComparator=new PersonComparator()
  if(personComparator.compare(p1,p2)>0) println(p1)
  else println(p2)
}
```

代码中直接使用的是 java.util.Comparator，即

```
public interface Comparator<T>
```

在 Scala 语言中使用时直接通过代码 class PersonComparator extends Comparator[Person]就可以，也即将 Java 语言中的泛型使用语法转换成 Scala 语言中的泛型使用语法即可。

13.2.2 Java 中使用 Scala 泛型

同样，Java 语言中也可使用 Scala 泛型，首先定义 Scala 类 Student 并指定其参数为泛型，具体代码如下：

```
package cn.scala.chapter13
import scala.beans.BeanProperty

//Student 类用泛型定义，成员变量 name 及 age 指定泛型参数
//并且用注解的方式生成 JavaBean 规范的 getter 方法
//因为是 val 的，所以只会生成 getter 方法
class Student[T,S](@BeanProperty val name:T,@BeanProperty val age:S){
```

}

在 Java 中使用 Student 类,其代码如下:

```java
package cn.scala.chapter13;
/**
 * Java 中使用 Scala 泛型代码示例
 */
public class ScalaGenericInJava {
    public static void main(String[] args){
        //指定 Student 类的两个泛型参数,name 为 String, age 为 Integer
        Student<String,Integer> student=new Student<String,Integer>("小李",18);
        //Scala 版本的 getter 方法
        System.out.println(student.name());
        //JavaBean 版本的 getter 方法
        System.out.println(student.getName());
    }
}
```

通过上述代码不难看出,Java 在使用 Scala 定义的泛型类时,与使用 Java 语言中定义的泛型类并没有实质性的差别。

还有一个值得注意的地方就是 Java 带通配符的泛型与 Scala 泛型间的互操作,定义 Java 类 JavaWildcardGeneric,具体代码如下:

```java
import java.util.ArrayList;
import java.util.List;
/**
 * Java 通配符泛型
 */
public class JavaWildcardGeneric {
    //Java 的通配符类型,可以接受任何类型
    public static List<?> getList(){
        List<String> listStr=new ArrayList<String>();
        listStr.add("Hadoop");
        listStr.add("Hive");
        return listStr;
    }
}
```

Scala 中并不支持?通配符,而是通过存在类型来表达这一语法,具体代码如下:

```
import java.util.List
import scala.collection.JavaConversions._
/**
 * 在 Scala 代码中使用 Java 通配符泛型代码演示
 */
class ScalaExistTypeToJavaWildcardGeneric {
  //采用 Scala 中的存在类型与 Java 中的通配符泛型进行互操作
  def printList(list: List[T] forSome {type T}):Unit={
    //因为我们引入了 import scala.collection.JavaConversions._
    //所以可以直接调用 foreach 方法
    list.foreach(println)
  }
  //上面的函数与下面的等同
  def printList2(list: List[_]):Unit={
    list.foreach(println)
  }
}
object Main extends App{
  val s=new ScalaExistTypeToJavaWildcardGeneric
  s.printList(JavaWildcardGeneric.getList)
  s.printList2(JavaWildcardGeneric.getList)
}
```

代码运行结果如下:

```
Hadoop
Hive
Hadoop
Hive
```

13.3　Scala trait 在 Java 中的使用

当 Scala trait 中的方法全部是抽象的时候，它与 Java 中的 interface 是等同的，下面的代码定义了一个 trait 并在该 trait 中定义了 4 个抽象方法：

```
//全部是抽象成员，与 java 的 interface 等同
trait MySQLDAO{
  def delete(id:String):Boolean
  def add(o:Any):Boolean
```

```
    def update(o:Any):Int
    def query(id:String):List[Any]
}
```

反编译后可以看到如下代码:

```
E:\IntellijIDEAWorkspace\out\production\ScalaProject\cn\scala\chapter13>javap -private -p MySQLDAO.class
Compiled from "MySQLDAO.scala"
public interface cn.scala.chapter13.MySQLDAO {
  public abstract boolean delete(java.lang.String);
  public abstract boolean add(java.lang.Object);
  public abstract int update(java.lang.Object);
  public abstract scala.collection.immutable.List<java.lang.Object> query(java.lang.String);
}
```

此时在 Java 代码中可以把 MySQLDAO 当作是一个普通的 interface 来使用，具体示例如下:

```
package cn.scala.chapter13;

import scala.collection.immutable.List;

/**
 * Trait MySQLDAO 中全部为抽象方法时,Java 代码中调用方式示例
 */
//直接通过 implements 关键字,看作是一个普通的 interface 来使用
class MySQLDAOImpl implements MySQLDAO{
    @Override
    public boolean delete(String id) {
        System.out.println("delete");
        return false;
    }

    @Override
    public boolean add(Object o) {
        System.out.println("add");
        return false;
    }
```

```java
    @Override
    public int update(Object o) {
        System.out.println("update");
        return 0;
    }

    @Override
    public List<Object> query(String id) {
        System.out.println("query");
        return null;
    }
}
public class TraitInJava {
    public static void main(String[] args) {
        MySQLDAO mySQLDAO=new MySQLDAOImpl();
        mySQLDAO.delete("100");
    }
}
```

但如果 MySQLDAO 定义了具体方法,那又要怎么样呢?见下面的示例代码:

```scala
package cn.scala.chapter13

//trait 中包含具体方法
trait MySQLDAO1{
  def delete(id:String):Boolean={
    println("delete")
    false
  }
  def add(o:Any):Boolean
  def update(o:Any):Int
  def query(id:String):List[Any]
}
```

此时如果还使用 implements 关键字则会出错,具体代码如下:

```java
class MySQLDAOImpl1 implements MySQLDAO1{
    @Override
    public boolean delete(String id) {
        //此时 super.delete 方法报如下错误
```

```java
//Error:(8, 21) java: 找不到符号
    //符号: 方法 delete(java.lang.String)
        return super.delete(id);
    }

    @Override
    public boolean add(Object o) {
        return false;
    }

    @Override
    public int update(Object o) {
        return 0;
    }

    @Override
    public List<Object> query(String id) {
        return null;
    }
}
```

有两种方法可以解决上述问题，一种是在 Scala 代码中对 MySQLDAO1 进行扩展并给出其他抽象方法的实现，然后再在 Java 代码中使用扩展后的类，另外一种方法是调用编译后生成的带有具体 delete 方法实现的 MySQLDAO1$class。首先看一下扩展 MySQLDAO1 来避免这一问题的具体示例：

```scala
//直接在 Scala 对 MySQLDAO1 进行扩展
class MySQLDAOImpl1 extends MySQLDAO1{
  override def delete(id: String): Boolean = super.delete(id)
  def add(o:Any):Boolean={println("add");true}
  def update(o:Any):Int={println("update");1}
  def query(id:String):List[Any]={println("query");null}
}
```

然后在 Java 中调用，如下所示：

```java
public class TraitInJava1 {
    public static void main(String[] args) {
        MySQLDAO1 mySQLDAO=new MySQLDAOImpl1();
        mySQLDAO.delete("100");
    }
}
```

上面的代码演示的是通过 Scala 类对 MySQLDAO1 进行 extends 操作来避免 trait 中带有具体方法时与 Java 进行互操作遇到的问题，还有一种方法就是在 Java 代码中调用编译后生成的带有具体 delete 方法实现的 MySQLDAO1$class，具体代码如下：

```java
package cn.scala.chapter13;

import scala.collection.immutable.List;

class MySQLDAOImpl1 implements MySQLDAO1{
    @Override
    public boolean delete(String id) {
     // MySQLDAO1 会生成两个字节码文件，分别是 MySQLDAO1$class.class 和 MySQLDAO1.class
        //MySQLDAO1$class.class 中有该方法的具体实现，直接调用即可
        if (MySQLDAO1$class.delete(this, id)) return true;
        else return false;
    }

    @Override
    public boolean add(Object o) {
        return false;
    }

    @Override
    public int update(Object o) {
        return 0;
    }

    @Override
    public List<Object> query(String id) {
        return null;
    }
}
public class TraitInJava1 {
    public static void main(String[] args) {
        MySQLDAO1 mySQLDAO=new MySQLDAOImpl1();
        mySQLDAO.delete("100");
    }
}
```

通过上述代码的演示可以看出，Scala 与 Java 进行互操作时，如果 Java 代码中要调用 Scala 中定义的 trait，此时需要特别注意的是 Scala trait 中定义的方法是否存在具体方法，如果不存在具体方法则可以直接作为 interface 使用，如果存在具体方法则需要做特殊处理。一种是在 Scala 中先对 trait 进行扩展然后再在 Java 代码中调用，另外一种便是直接调用 trait 编译后生成的 ***$class 中的具体方法，从简洁性和直观性来讲，建议先在 Scala 中对 trait 进行 extends，然后再在 Java 中调用扩展后的类。

13.4 Scala 与 Java 异常处理互操作

在前面讲 Scala 类的时候提到，如果想要生成 Java 风格的 getter、setter 方法，需要用 @reflect.BeanProperty 对类成员变量进行注解，这里我们对 Annotation 的作用再进行讲解。我们知道，Scala 要捕获 Java 抛出的异常可以用下列代码：

```scala
import java.io.File
/**
 *Scala 捕获 Java 代码抛出的异常
 */
object ScalaExceptionDemo extends App{
  val file: File = new File("a.txt")
  if (!file.exists) {
    try {
      file.createNewFile
    }
    catch {
      //通过模式匹配来实现异常处理
      case e: IOException => {
        e.printStackTrace
      }
    }
  }
}
```

那在 Java 代码中又怎么去捕获 Scala 代码抛出的异常呢？捕获方法就是通过 annotation 来实现，具体代码如下：

```scala
//Scala 代码
class ScalaThrower {
  //Scala 利用注解@throws 声明抛出异常
  @throws(classOf[Exception])
  def exceptionThrower {
    throw new Exception("Exception!")
  }
}
```

```java
//Java 代码
//Java 中调用 ScalaThrower(Scala 类), 然后捕获其抛出的异常
public class JavaCatchScalaThrowerException {
    public static void main(String[] args){
        ScalaThrower st=new ScalaThrower();
        try{
            st.exceptionThrower();
        }catch (Exception e){
            e.printStackTrace();
        }
    }
}
```

小 结

本章对 Scala 与 Java 语言间的互操作问题进行了介绍，着重介绍了 Java 集合与 Scala 集合间的互操作问题、Java 泛型与 Scala 泛型间的互操作问题，对 Java 语言在使用 Scala trait 时的互操作问题进行了详细介绍，最后介绍了 Scala 语言与 Java 语言在异常处理方面的互操作。

参考文献

[1] Martin Odersky，Lex Spoon，Bill Venners. Programming in Scala. California: Artima Inc，2008.

[2] Joshua D.Suereth.Scala Cookbook. New York California: Manning Publications，2012.

[3] Cay S.Horstmann. Scala for the Impatient.Boston: Addison-Wesley Professional，2012.

[4] Joshua D.Suereth.Scala in Depth.New York: Manning Publications，2012.

[5] Nilanjan Raychaudhuri.Scala in Action. New York: Manning Publications，2013.

[6] Munish K.Gupta.Akka Essentials. Birmingham: Packt Publishing，2012.

[7] http://doc.akka.io/docs/akka/snapshot/scala